DARK ECOLOGY

THE WELLEK LIBRARY LECTURES IN CRITICAL THEORY

The Wellek Library Lectures in Critical Theory are given annually at the University of California, Irvine, under the auspices of the Critical Theory Institute. The following lectures were given in May 2014

THE CRITICAL THEORY INSTITUTE
Gabriele Schwab, Director

DARK ECOLOGY

For a Logic of Future Coexistence

TIMOTHY MORTON

Columbia University Press
New York

Columbia University Press
Publishers Since 1893
New York Chichester, West Sussex
cup.columbia.edu

Library of Congress Cataloging-in-Publication Data

Names: Morton, Timothy, 1968– author.
Title: Dark ecology: for a logic of future coexistence / Timothy Morton.
Description: New York : Columbia University Press, 2016. |
Series: Wellek Library lectures in critical theory | Includes bibliographical references and index
Identifiers: LCCN 2015026796 | ISBN 9780231177528 (cloth : alk. paper) |
ISBN 9780231541367 (e-book)
Subjects: LCSH: Nature—Effect of human beings on —Philosophy. | Human beings—Effect of environment on—Philosophy. | Human ecology—Philosophy.
| Naturalness (Environmental sciences)
Classification: LCC GF75 .M685 2016 | DDC.2—dc23
LC record available at http://lccn.loc.gov/2015026796

Columbia University Press books are printed on
permanent and durable acid-free paper.
This book is printed on paper with recycled content.
Printed in the United States of America

Cover design by Julia Kushnirsky
Cover illustration by Hannah Stouffer

For Allan

It was surprising how pure the sense of loss was—in a sense it's because nonhumans don't have the same mediation with humans. I mean, you know your grandma or whoever is sick and that there is a hospital, there is a timeline of some kind. That and the fact that the "outside" (I should know this by now) is actually the (human) inside, so it's strangely like losing a child to war. It's a war zone against nonhumans.

He was hit by the mail truck, the new delivery person having a habit of driving up the driveway. The worst aspect was that he tried to crawl back in with a smashed neck and head, so I found him right outside the cat door, still warm yet with rigor mortis. We buried him like an Egyptian with his favorite things and did a Buddhist death ritual right away. For the next few days we were totally rigid with depression, which slowly liquefied.

He wasn't murdered—though for a moment the obvious blunt force trauma to his neck looked very like it. One of my friends had a cat who was indeed killed by some psychopath who showed her the cat's body in his freezer. Nevertheless Allan Whiskersworth was killed by humans in a "friendly fire" "collateral damage" sort of way. Cats weirdly symbolize the ambiguous border between agricultural logistics and its (impossible to demarcate) outside. I mean we don't let dogs just wander about. It's as if we want to use cats to prove to ourselves that there is a Nature. Allan was very happy bristling among the grasses and talking to his friend, the gray cat. He lived a Neil Young sort of life (burning out) and died at only two years old. I've always liked Lennon's response that he'd much rather fade away (and look what happened to him).

Right after his death the Charon-like gray cat came to visit, never before or since.

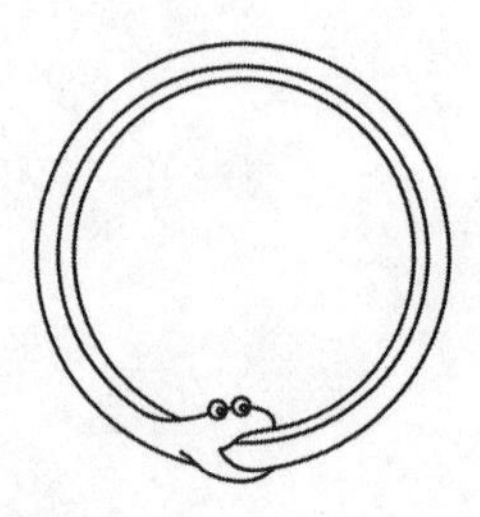

Progress means: humanity emerges from its spellbound state no longer under the spell of progress as well, itself nature, by becoming aware of its own indigenousness to nature and by halting the mastery over nature through which nature continues its mastery.

Theodor Adorno

Dark is dangerous. You can't see anything in the dark, you're afraid. Don't move, you might fall. Most of all, don't go into the forest. And so we have internalized this horror of the dark.

Hélène Cixous

CONTENTS

ACKNOWLEDGMENTS

This book is a version of the Wellek Lectures that I gave in May 2014 at the University of California, Irvine. My thanks go to Georges Van Den Abeele, dean of humanities, for having invited me and for being a huge inspiration on this project. Georges and a host of scholars kindly donated their intellects for three days at UC Irvine, near the set of *Conquest of Planet of the Apes*, including Ackbar Abbas, Ellen Burt, David Theo Goldberg, Martin Harries, Andrea Henderson, Jayne Lewis, Julia Lupton, Glen Mimura, Beryl Schlossman, Gabriele Schwab, Michael Szalay, and Christopher Tomlins.

Thanks so much to my editor at Columbia University Press, Jennifer Crewe. Her intellectual generosity and accurate perception were invaluable. And thanks to my dean, Nicholas Shumway, whose unfailing support and encouragement has been such a gift over the last three years.

A great deal of the research that went into this book would not have been possible without the steady and generous correspondence of Dirk Felleman, Jarrod Fowler, and Cliff Gerrish. Thank you, friends. And thank you to my tireless research assistants, Jade Hagan and Mallory Pladus.

In the fall of 2014 a sound art collective called Sonic Acts started the "Dark Ecology" art project, inspired by my thoughts on the topic at hand since 2004. My thanks to all those who have invited me to talk at numerous events and witness the iron smelter at Nikel in

Arctic Russia.I am particularly grateful to Arie Altena, Lucas van der Velden, and Annette Wolfsberger. For the last three years, Katherine Behar's invitations to participate in the Object-Oriented Feminism panels at the Society for the Study of Literature, Science, and the Arts have been invaluable.

A host of artists, curators, and scholars have contributed invaluable things to this project. Thoughts are interactive, and Q&A sessions are my lab. For their incredible help and kindness, my heartfelt thanks also go to Mirna Belina, Klaus Biesenbach, Dominic Boyer, Joseph Campana, Martin Clark, Thomas Claviez, Jeffrey Cohen, Tom Cohen, Claire Colebrook, Dipesh Chakrabarty, Lowell Duckert, Olafur Eliasson, William Elliott, Ine Gevers, Alejandro Ghersi, Jón and Jøga Gnarr, Rune Graulund, Richard Grossinger, Björk Guðmundsdóttir, Graham Harman, Erich Hörl, Cymene Howe, Serenella Iovino, Chuck Johnson, Toby Kamps, Douglas Kahn, Jeffrey Kripal, Jae Rhim Lee, Charles Long, Frenchy Lunning, Andrea Mellard, James Merry, Hilde Methi, Julia Nuessiein, Hans Ulrich Obrist, Henk Oosterling, Serpil Opperman, John Palmesino, Heather Pesanti, Karen Pinkus, Albert Pope, Ann Sofi Rönnskog, Judith Roof, Nicholas Royle, Miljohn Ruperto, Maria Rusinovskaya, Roy Sellars, Rhoda Seplowitz, Christopher Schaberg, Emilija Škarnulyte, Haim Steinbach, Thom van Dooren, Leslie Uppinghouse, Gediminas Urbonas, Lynn Voskuil, Laura Walls, Cary Wolfe, Carolyn Wyatt, and Marina Zurkow.

The cute ouroboros was drawn for me by the magical Ian Bogost.

DARK ECOLOGY

BEGINNING AFTER THE END

There are thoughts we can anticipate, glimpsed in the distance along existing thought pathways.
This is a future that is simply the present, stretched out further.
There is not-yet-thought that never arrives—yet here we are thinking it in the paradoxical flicker of this very sentence.
If we want thought different from the present—if we want to change the present—then thought must be aware of this kind of future.
It is not a future into which we can progress.
This future is unthinkable. Yet here we are, thinking it.
Coexisting, we are thinking future coexistence. Predicting it and more: keeping the unpredictable one open.
Yet such a future, the open future, has become taboo.
Because it is real, yet beyond concept.
Because it is *weird*.
Art is thought from the future. Thought we cannot explicitly think at present. Thought we may not think or speak at all.
If we want thought different from the present, then thought must veer toward art.
To be a thing at all—a rock, a lizard, a human—is to be in a twist.
How thought longs to twist and turn like the serpent poetry!

Or is art veering toward thought? Does it ever arrive?
The threads of fate have tied our tongues.
Tongue twisters inclined towards nonsense.
Logic includes nonsense as long as it can tell the truth.
The logic of nonsense.
The needle skipped the groove of the present.
Into this dark forest you have already turned.
I take *present* to mean *for the last twelve thousand years*. A butterfly kiss of geological time.

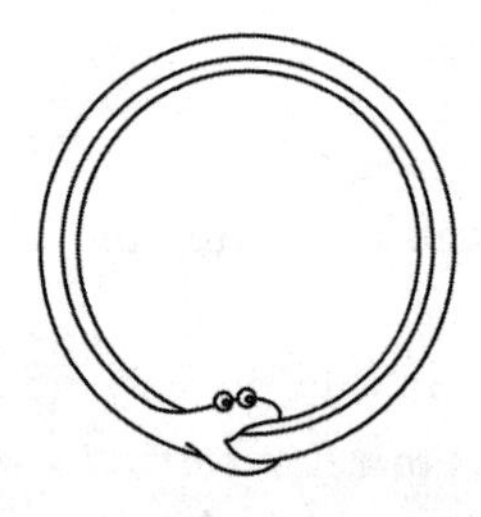

THE FIRST THREAD

Each outcry from the hunted Hare
A fibre from the Brain does tear

What is happening?

The field had already been "opened" . . . a lane a few feet wide had been hand-cut through the wheat along the whole circumference of the field for the first passage of the horses and machine.

Two groups, one of men and lads, the other of women, had come down the lane just at the hour when the shadows of the eastern hedge-top struck the west hedge midway, so that the heads of the groups were enjoying sunrise while their feet were still in the dawn . . .

Presently there arose from within a ticking like the love-making of the grasshopper. The machine had begun, and a moving concatenation of three horses and the aforesaid long rickety machine was visible over the gate. . . . Along one side of the field the whole wain went, the arms of the mechanical reaper revolving slowly . . .

The narrow lane of stubble encompassing the field grew wider with each circuit, and the standing corn was reduced to a smaller area as the morning wore on. Rabbits, hares, snakes,

> rats, mice, retreated inwards as into a fastness, unaware of the ephemeral nature of their refuge, and of the doom that awaited them later in the day when, their covert shrinking to a more and more horrible narrowness, they were huddled together . . . till the last few yards of upright wheat fell also under the teeth of the unerring reaper, and they were every one put to death by the sticks and stones of the harvesters.
>
> The reaping-machine left the fallen corn behind it in little heaps, each heap being of the quantity for a sheaf; and upon these the active binders in the rear laid their hands—mainly women, but some of them men . . .
>
> [The women] were the most interesting of this company of binders, by reason of the charm which is acquired by woman when she becomes part and parcel of outdoor nature. . . . A field-man is a personality afield; a field-woman is a portion of the field; she had somehow lost her own margin . . . and assimilated herself with it.
>
> . . . There was one wearing a pale pink jacket . . .
>
> Her binding proceeds with clock-like monotony. From the sheaf last finished she draws a handful of ears, patting their tips with her left palm to bring them even. Then, stooping low, she moves forward, gathering the corn with both hands against her knees, and pushing her left gloved hand under the bundle to meet the right on the other side, holding the corn in an embrace like that of a lover. She brings the ends of the bond together, and kneels on the sheaf while she ties it, beating back her skirts now and then when lifted by the breeze. A bit of her naked arm is visible . . . and as the day wears on its feminine smoothness becomes scarified by the stubble and bleeds.[1]

It's the machine age, yet uncannily it isn't: it's fields and wheat. Or are the fields already a kind of machine? People appear as machine-like components, legs, clothing, arms, and hands moving. Tess of the

D'Urbervilles, a fictional farming girl from 1891, appears as if she were a piece of a gigantic device, yet she also as a human individual, exemplifying a weird contradiction between being and appearing.[2] Seeing this contradiction, enabled by the machination of steam engines and Kantian code, forces us to think a far, far older machination, still churning. A twelve-thousand-year structure, a structure that seems so real we call it Nature. The slowest and perhaps most effective weapon of mass destruction yet devised.[3]

What is dark ecology?[4] It is ecological awareness, dark-depressing. Yet ecological awareness is also dark-uncanny. And strangely it is dark-sweet. Nihilism is always number one in the charts these days. We usually don't get past the first darkness, and that's if we even care. In this book we are going to try to get to the third darkness, the sweet one, through the second darkness, the uncanny one. Do not be afraid.

What thinks dark ecology? *Ecognosis*, a riddle. Ecognosis is like knowing, but more like letting be known. It is something like coexisting. It is like becoming accustomed to something strange, yet it is also becoming accustomed to strangeness that doesn't become less strange through acclimation. Ecognosis is like a knowing that knows itself. Knowing in a loop—a *weird* knowing. *Weird* from the Old Norse *urth*, meaning twisted, *in a loop*.[5] The Norns entwine the web of fate with itself; Urðr is one of the Norns.[6] The term *weird* can mean *causal:* the winding of the spool of fate. The less well-known noun *weird* means *destiny* or *magical power* and, by extension, the wielders of that power, the Fates or Norns.[7] In this sense *weird* is connected with *worth*, not the noun but the verb, which has to do with *happening* or *becoming*.[8]

Weird: a turn or twist or loop, a turn of events. The milk turned sour. She had a funny turn. That weather was a strange turn-up for the book. Yet *weird* can also mean *strange of appearance*.[9] That storm cloud looks so weird. She is acting weird. The milk smells weird. Global weirding.

In the term *weird* there flickers a dark pathway between causality and the aesthetic dimension, between doing and appearing, a pathway that dominant Western philosophy has blocked and suppressed. We shall be traveling down this pathway because it provides an exit route from the machinelike functioning of Tess's field. Now the thing about seeming is that seeming is never quite as it seems. *Dark Ecology* is going to argue that appearance is always strange. We discern yet another pathway, a route between the term *weird* and the term *faerie*.[10] *Faerie* also comes from a word for *fate* and suggests a "supernatural" illusion-like magical appearance as well as a kind of "unearthly" realm:

> weird << *urth* (Norse) = Norn = twisting fate = *fatum* (Latin) >> fay >> faerie

Though the web of fate is so often invoked in tragedy, that default agricultural mode, words such as *weird* and *faerie* evoke the animistic world within the concept of the web of fate itself. The dark shimmering of *faerie* within *fate* is a symptom of what *Dark Ecology* is going to attempt. We are going to try to see how we Mesopotamians have never left the Dreaming. So little have we moved that even when we thought we were awakening we had simply gathered more tools for understanding that this was in fact a lucid dream, even better than before.

Weird weirdness. Ecological awareness is weird: it has a twisted, looping form. Since there is no limit to the scope of ecological beings (biosphere, solar system), we can infer that all things have a loop form. Ecological awareness is a loop because human interference has a loop form, because ecological and biological systems are loops. And ultimately this is because to exist at all is to assume the form of a loop. The loop form of beings means we live in a universe of finitude and fragility, a world in which objects are suffused with and surrounded by mysterious hermeneutical clouds of unknowing. It means that the politics of coexistence are always contingent, brittle, and flawed, so that in the thinking of interdependence at least one being must be missing. Ecognostic jigsaws are never complete.

What kind of weirdness are we talking about? Weird weirdness. Weird means *strange of appearance*; weirdness means the *turning* of causality. Let's focus this idea by thinking about the many kinds of ecological loops. There are *positive feedback loops* that escalate the potency of the system in which they are operating. Antibiotics versus bacteria. Farmers versus soil, creating the Dust Bowl in the Midwestern United States in the 1930s. Such loops are common in human "command and control" approaches to environmental management, and they result in damage to ecosystems.[11] Some of them are unintended: consider the decimation of bees in the second decade of the twenty-first century brought on by the use of pesticides that drastically curtail pollination.[12] Such unintended consequences are *weirdly weird* in the sense that they are uncanny, unexpected fallout from the myth of progress: for every seeming forward motion of the drill bit there is a backward gyration, an asymmetrical contrary motion.

Then there are the *negative feedback loops* that cool down the intensity of positive feedback loops. Think of thermostats and James Lovelock's Gaia. There are *phasing loops*. We encounter them in beings such as global warming, beings that are temporally smeared in such a way that they come in and out of phase with human temporality. (This book is going to call it *global warming*, not *climate change*.)[13]

Yet there is another loop, the dark-ecological loop: a *strange loop*. A strange loop is one in which two levels that appear utterly separate flip into one another. Consider the dichotomy between moving and being still. In Lewis Carroll's haunting story, Alice tries to leave the Looking Glass House. She sets off through the front garden, yet she finds herself returning to the front door via that very movement.[14] A strange loop is weirdly weird: a turn of events that has an uncanny appearance. And this defines emerging ecological awareness occurring to "civilized" people at this moment.

Two kinds of weird: a turning and a strange appearing, and a third kind, the weird gap between the two. The Anthropocene names two levels we usually think are distinct: geology and humanity. Since the late eighteenth century humans have been depositing layers of carbon in Earth's crust. In 1945 there occurred the *Great Acceleration* of the Anthropocene, marked by a huge data spike in the graph of human involvement in Earth systems. The Anthropocene binds together human history and geological time in a strange loop, weirdly weird. Consider how personal this can get. There you were, shoveling coal into your steam engine, that great invention patented in 1784 that Marx hails as the driver of industrial capitalism. The very same machine that Paul Crutzen and Eugene Stoermer hail as the instigator of the Anthropocene.[15] The year 1784 is not the earliest date for a steam engine patent, but the language of the 1784 patent describes the engine as a general-purpose machine that can be connected to any other machine in order to power it. This general-purpose quality enables the industrial *turn*.

There you are, turning the ignition of your car. And it creeps up on you. You are a member of a massively distributed thing. This thing is called *species*. Yet the difference between the weirdness of my ignition key twist and the weirdness of being a member of the human species is itself weird. Every time I start my car or steam engine I don't mean to harm Earth, let alone cause the Sixth Mass Extinction Event in the four-and-a-half billion-year history of life on this planet.[16] (Disturbingly, the most severe extinction so far in Earth history, the End-Permian Extinction, was very likely caused by global warming.)[17] Furthermore, I'm not harming Earth! My key turning is statistically meaningless. In an individual sense this turn isn't weird at all.

But go up a level and something very strange happens. When I scale up these actions to include billions of key turnings and billions of coal shovelings, harm to Earth is precisely what is happening. I am responsible as a member of this species for the Anthropocene. Of course I am formally responsible to the extent that I understand global warming. That's all you actually need to be responsible for something. You understand that this truck is going to hit that man?

You are responsible for that man. Yet in this case formal responsibility is strongly reinforced by causal responsibility. I am the criminal. And I discover this via scientific forensics. Just like in noir fiction: I'm the detective *and* the criminal! I'm a person. I'm also part of an entity that is now *a geophysical force on a planetary scale*.[18]

The darkness of ecological awareness is the darkness of noir, which is a strange loop: the detective is a criminal. In a strong version of noir the narrator is implicated in the story: two levels that normally don't cross, that some believe *structurally can't cross*. We "civilized" people, we Mesopotamians, are the narrators of our destiny. Ecological awareness is that moment at which these narrators find out that they are the tragic criminal.

And what an astonishing reversal, what a twist or as Aristotle says about tragic downfalls, what a *peripeteia*—which technically is the moment at which a runner turns around a post in an ancient Greek stadium. A turn, a twist—something weird. What an astounding upsetting of our modern and postmodern fictions about the human and "the West." There are so many fictions that enumerating them all would take too long: just consider a central one having to do with our thoughts about where we live, the planet we inhabit. We have been telling ourselves that homogeneous empty "space" has conquered localized, particular "place." We are either the kind of person who thinks that the category of place is a quaint antique or we are the kind of person who thinks that the category is worth preserving because it *is* antique.[19] In a certain way, we are the same kind of person.

Many have pronounced the death of place since the 1970s. In literary studies the announcement has gone hand in hand with the language of textuality versus speech.[20] Our habitual talk pits speech (presence, villages, the organic, slow time, traditions) against textuality (dissolution, speed, modern, and postmodern technocultures). Yet the coordinates are terribly out of date. In a twist no one saw coming (because we weren't looking outside the human), space has by no means conquered place. That postmodern meme was simply a late symptom of the modern myth of transcending one's material conditions.

Exactly the opposite has occurred. From the standpoint of the genuinely post-modern ecological era, what has collapsed is (the fantasy of empty, smooth) *space*.[21] "Space" has revealed itself as the convenient fiction of white Western imperialist humans, just as relativity theory revealed Euclidean geometry to be a small human-flavored region of a much more liquid Gaussian spacetime. The Euclidean concept that space is a container with straight lines is good enough to be getting on with if you want to voyage around the coast of Africa to reach the Spice Islands. *Space* in this sense has collapsed, and *place* has emerged in its truly monstrous uncanny dimension, which is to say its nonhuman dimension. How? Now that the globalization dust has settled and the global warming data is in, we humans find ourselves on a very specific planet with a specific biosphere. It's not Mars. It is planet Earth. Our sense of planet is not a cosmopolitan rush but rather the uncanny feeling that there are all kinds of places at all kinds of scale: dinner table, house, street, neighborhood, Earth, biosphere, ecosystem, city, bioregion, country, tectonic plate. Moreover and perhaps more significantly: bird's nest, beaver's dam, spider web, whale migration pathway, wolf territory, bacterial microbiome. And these places, as in the concept of spacetime, are inextricably bound up with different kinds of timescale: dinner party, family generation, evolution, climate, (human) "world history," DNA, lifetime, vacation, geology; and again the time of wolves, the time of whales, the time of bacteria.

So many intersecting places, so many scales, so many nonhumans. Place now has nothing to do with good old reliable constancy. What has dissolved is the idea of *constant presence*: the myth that something is real insofar as it is consistently, constantly "there." The concept *space* was always a constant-presencing machine for making things appear consistent and solid, to make them easier to colonize, enslave, and plunder. Constant presence was part of an anthropocentric colonization protocol. The planetary awareness vaguely imagined by white Western humans in fantasies about Spice Islands and global trade is now upon us, and it has nothing to do with the rush of deterritorialization, of finding oneself unbound and unhinged.[22] It is almost the opposite. One finds oneself on the insides of much bigger *places* than those constituted by humans. Whose place is it anyway?

It is *space* that has turned out to be the anthropocentric concept, now that we are able to think it without a myth of constant presence. Celebrations of deracination and nostalgia for the old ways are both fictional. It is as obvious to any indigenous culture as it now is to anyone with data sets about global warming that these were stories white Westerners were telling themselves, two sides of the same story in fact. The ecological era is the revenge of place, but it's not your grandfather's place. This isn't some organic village we find ourselves in, nor indeed a city-state surrounded by fields.

Place has a strange loop form because place deeply involves time. Place doesn't stay still, but bends and twists: place *is* a twist you can't iron out of the fabric of things. When you are near your destination you can sometimes feel quite disoriented. You may enhance the magnification on Google Maps to make sure you are really there. The local is far from the totally known or knowable. It is familiar, which also means that it is uncanny (German, *heimisch*, "familiar" and "unfamiliar," "intimate" and "monstrous" at the same time). Nearness does not mean obviousness: just ask someone looking at a dust mite down a scanning electron microscope. When massive entities such as the human species and global warming become thinkable, they grow near. They are so massively distributed we can't directly grasp them empirically. We vaguely sense them out of the corner of our eye while seeing the data in the center of our vision. These "hyperobjects" remind us that *the local is in fact the uncanny*.[23] Space evaporates. The nice clean box has melted. We are living on a Gaussian sphere where parallel lines do indeed meet. The empty void of space and the rush of infinity have been unmasked as parochial paradigms.

The holism in which the whole is greater than sum of its parts depends on some (false) concept of smooth, homogeneous universality or space or infinity. It depends, in short, on a Euclidean anthropocentric geometry. Since they do not fit into the quaint category of space, what hyperobjects reveal to us humans is that the whole is always weirdly *less* than the sum of its parts. Take the new cities springing up, megacities such as Houston. For architects and urban planners, megacities are hard to conceptualize: where do they start and stop? Can one even point to them in a straightforward way? And isn't it strange that entities so obviously gigantic and so colossally influential on their surroundings and economies worldwide should be so hard to point to? The fact that we can't point to megacities is deeply because we've been looking in the wrong place for wholes. We keep wondering when the pieces will add up to something much greater. But now that we are truly aware of the global (as in global warming), we know that a megacity is a place among places, that is to say a finitude that contains all kinds of other finitudes, fragile and contingent. Like Doctor Who's time-and-space-traveling, the TARDIS, it's bigger on the inside than it is on the outside. Places contain multitudes.

And this has a retroactive corrosive effect. There never was a constantly present, easy to identify whole, because there was never a general, homogeneous space box. When you look back at the earliest city-states such as Damascus, you end up seeing the same thing as the megacities: uncertain boundaries, centers that never quite establish themselves as centers . . . why? Is it just a case of historical projection? Or is it rather because the city and the city-state are major symptoms of a gigantic elephant in the room, the elephant that eventually caused globalization, with its global warming and its ironic by-product, awareness of global warming?

An inconvenient Anthropocene. Not all of us are ready to feel sufficiently creeped out. Not a day goes by recently without some humanities scholars becoming quite exercised about the term *Anthropocene*, which has arisen at a most inconvenient moment. *Anthropocene* might sound to posthumanists like an anthropocentric symptom of a sclerotic era. Others may readily recall the close of Foucault's *The Order of Things*: "man" is like a face drawn in sand, eventually wiped away by the ocean tides.[24] What a weirdly prescient image of global warming, with its rising sea levels and underwater government meetings.[25] But how ironic—how strangely looped. There we were, happily getting on with the obliteration business, when *Anthropocene* showed up. The human returns at a geological level far deeper than sand. Give a posthumanist a break! This is also an inconvenient truth for those convinced that any hint of talk about reality smacks of reactionary fantasy, a bullying, know-nothing kick of a pebble.

The Sixth Mass Extinction Event: caused by the Anthropocene, caused by humans. Not jellyfish; not dolphins; not coral. The panic seems more than a little disingenuous given what we know about global warming, and given what we humanities scholars think we like to say about the role of humans in creating it, as opposed to, say, Pat Robertson or UKIP (the UK Independence Party). A Fredric Jameson might smile somewhat ruefully at the dialectic of scholars refusing the very concept of reality and big pictures, while global megacorporations frack in their backyards.

The ocean's silver screen. The trouble with global warming is that one can't just palm it off on a particular group of humans or insist that the Sixth Mass Extinction Event is just another construct. The humanities have persistently argued, via Foucault via Heidegger or Nietzsche or Marx via Hegel via Kant, that there are no accessible things in themselves, only thing-positings or thingings of Dasein or thing discourses or things posited by the history of spirit or will or (human) economic relations. Only things insofar as they correlate to some version of the (human) subject, which is why this thinking is called correlationist.[26] But the screen on which these correlations are projected isn't blank after all. It consists of unique, discrete entities with a "life" of their own no matter whether a (human) subject has opened the epistemological refrigerator door to check them. Some entities violently treated as blank screens are overwhelming human being itself, as what the insurance industry calls *acts of God* turn out to be acts of humans as a geophysical force.

Foucault's face in the sand depicts the regime of power-knowledge that begins in 1800, another strange turn of events. Eighteen hundred is the moment of the steam engine, engine of the Anthropocene. Eighteen hundred is also the moment of Hume and Kant, who inaugurated correlationism. Hume argued that cause and effect were mental constructs based on interpretations of data: hence the statistical methods of modern science. Which is why global warming deniers and tobacco companies are able to say, with something like a straight face, that "no one has ever proved" that humans caused global warming or that smoking causes cancer.

In the same way a post-Humean person can't claim that this bullet she is going to fire into my head at point-blank range is going to kill me. She can say that it's 99.9 percent likely, which is actually *better* since saying so relies only on data, not on metaphysical factoids culled from Aristotelian arguments about final causes. Thus the Intergovernmental Panel on Climate Change (IPCC) makes it more and more clear that humans have caused global warming, but they need to express this as a statistic: as I'm writing it's at 97 percent.[27] Which leaves an out for conservatives who like to deny global warming by going, "Look at this snowball, so there's no global warming at all!" In addition to denying global warming, denials involving snowballs are denying the only causality theories that make sense to us.

How I learned to stop worrying and love the term "Anthropocene." Let's examine the modes of Anthropocene denial. First, the claim of colonialism: the Anthropocene is the product of Western humans, mostly Americans. It unfairly lumps together the whole human race.

Although the desire for it first emerged in America, it turns out everyone wants air conditioning. On this issue I am in accord with Dipesh Chakrabarty, who had the courage to name the concept *species* on which the concept *Anthropocene* depends.[28] Likewise obesity isn't simply American. Americans are not like aspartame, ruining the natural sweetness of other humans. The deep reason why is that at no point in history did any human straightforwardly *need* something. Desire is logically prior to whatever "need" is, histories of consumerism notwithstanding, histories that tend to repeat Fall narratives not unrelated to the normal (and unhelpful) ways we think ecology: "First we needed things, then at point *x* we wanted things, and that put us into an evil loop." We think of loops as sin. But loops aren't sinful. There was no Fall, unless you believe in the Mesopotamian logic that eventually created global warming. There was no transition from "needing" to "wanting." Neanderthals would have loved Coca-Cola Zero.[29]

Secondly, racism. The user of *Anthropocene* is saying that humans as a race are responsible, and while this really means *white* humans, whites go unmarked.

There is such a thing as the human. But *human* need not be something that is ontically given: we can't see it or touch it or designate it as present in some way (as whiteness or not-blackness et cetera). There is no obvious, constantly present positive content to the human. So *Anthropocene* isn't racist. Racism exists when one fills in the gap between what one can see (beings starting engines and shoveling coal) and what this human thing is: the human considered as a species, namely as a hyperobject, a massively distributed physical entity of which I am and am not a member, simultaneously. (We'll see how there are Darwinian, phenomenological, and logical reasons for this violation of the "Law" of Noncontradiction). The racist effectively erases the gap, implicitly reacting against what Hume and Kant did to reality. Since their age we have thought it sensible that there is some kind of irreducible rift between what a thing is and how it appears, such that science handles data, not actual things.

Copyright control. I am myself a correlationist, by which I mean that I accept Kant's basic argument that when I try to find the thing in itself, what I find are thing data, not the thing in itself. And I grasp that data in such a way that a thing does not (meaningfully) exist (for me) outside the way I (or history or economic relations or will or Dasein) correlate that data. I believe that there is a drastic *finitude* that restricts my access to things in themselves. The finitude is drastic because it is irreducible. I can't bust through it. This marks the difference between some speculative realists, who think you can puncture the finitude and enter a world of direct access, for instance via science, and those who don't think so, for instance the object-oriented ontologists.

Object-oriented ontology, or OOO, developed from a deep consideration of the implications of Martin Heidegger's version of modern Kantian correlationism. These implications would have seemed bizarre to Kant and Heidegger themselves, who in their different ways (transcendental idealism and fascism) tried to contain the explosive vision that their thinking unleashed. Ontology doesn't tell you exactly what exists but *how* things exist. If things exist, they exist in *this* way rather than *that*. Object-oriented ontology holds that things exist in a profoundly "withdrawn" way: they cannot be splayed open and totally grasped by anything whatsoever, including themselves. You can't know a thing fully by thinking it or by eating it or by measuring it or by painting it . . . This means that the way things affect one another (causality) cannot be direct (mechanical), but rather indirect or vicarious: causality is aesthetic. As strange as this sounds, the idea that causality is aesthetic is congruent with the most powerful causality theories (the Humean ones), and the most powerful theories of causality in physical science: relativity theory and (to an even greater extent) quantum theory. In a way that profoundly differs from the demystification most popular in humanistic accounts of culture, politics, and philosophy (and so on), OOO believes that reality is *mysterious* and *magical*, because beings withdraw and because beings influence each other aesthetically, which is to say at a distance.[30]

If ecological culture and politics is about "the reenchantment of the world" as they say, then something like OOO could be highly desirable. In particular, the way in which OOO doesn't reject modern science and philosophy, but rather proceeds from them and somehow finds magic that way, is valuable indeed. We will be thinking through the ecological implications of the OOO view throughout *Dark Ecology*.

Finitude is the term that describes a world in which entities "withdraw" from direct access. Every kind of access—a philosopher thinking about a stone, a scientist smashing a particle, a farmer watering a tree—is profoundly limited and incomplete. And every type of *nonhuman* access—a thrush smashing a snail shell against a stone, an electron interacting with a photon, a tree absorbing water—is also profoundly limited. Kant was the philosopher who argued for this finitude, at least when it came to how *humans* access things. I don't believe that the finitude of the human-world correlate is incorrect. It can't be ripped open, even by something as seemingly sharp as mathematics.[31] When I mathematize a thing, there I am, mathematizing it—measuring it, for instance. Why this is so different a form of access than eating it or using it to paper my room is uncertain. The gap between the human and everything else can't be filled in, as racism tries to do.

There is a tactic we could adopt, a tactic deeply congruent with ecological politics. Kant grounded Hume's argument in synthetic judgments a priori in a transcendental subject (not "little me," the one I can see and touch). Only a correlator such as a (human) subject makes reality real. At the very moment at which philosophy says you can't directly access the real, humans are drilling down ever deeper into it, and the two phenomena are deeply, weirdly intertwined. Correlationism is true, but disastrous if restricted to humans only. Possibly more of a disaster than treating things as lumps is treating them as *blank lumps* we can format as we wish. How to proceed? We should merely *release the anthropocentric copyright control on correlationism*, allowing nonhumans like fish (and perhaps even fish forks) the fun of not being able to access the in-itself.

On this view, whether the thing in itself becomes fish food or human food or something a human can measure, the thing remains in excess of those forms of access, and there is no intrinsic superiority of human ways of accessing the thing. This is the basic premise of object-oriented ontology: Kant was correct, but his anthropocentrism prevented him from seeing the most interesting aspects of his theory. We will see that these aspects could have a profound influence on the way we think the logic of future coexistence.

Very well, says the hesitant humanist. *Anthropocene* may not be colonialist or racist, but surely it must be a blatant example of speciesism? Isn't the term claiming that humans are special and different, unique in having created it?

Humans and not dolphins invented steam engines and drilled for oil. But this isn't a sufficient reason to suppose them special. Etymology notwithstanding, *species* and *specialness* are extremely different. Just ask Darwin. Unfortunately he had no recourse to emoticons, for if his masterwork's title had contained a wink emoticon at its end, he could have said it succinctly: there are no species—and yet there are species! And they have no origin—and yet they do! A human is made up of nonhuman components and is directly related to nonhumans. Lungs are evolved swim bladders. Yet a human is not a fish.[32] A swim bladder, from which lungs derive, is not a lung in waiting. There is nothing remotely lunglike about it.[33] Let alone my bacterial microbiome: there are more bacteria in "me" than "human" components. A lifeform is what Derrida calls *arrivant* or what I call *strange stranger*: it is itself yet uncannily not itself at the same time.[34] Contemporary science allows us to think species not as absolutely nonexistent, but as floating, spectral entities that are not directly, constantly present. *Spectral* is in some senses cognate with *species*.

The Darwinian concept of species is precisely not the Aristotelian one where you can tell teleologically what species are for: ducks are for swimming, Greeks are for enslaving barbarians . . . Since *species* in this sense fails to coincide with me, an actual human being as opposed to a pencil or a duck, the concept of species isn't speciesist. Like the racist, the speciesist fills out the gap between phenomenon and thing with a special paste: the fantasy of an easy-to-identify content. That sort of content is what one is incapable of seeing, yet there are ducks and spoonbills, which are not humans.

The seemingly anachronistic and dangerous concept *species* appears superficially easy to think: *Sesame Street* ("We Are All Earthlings") conveys it.[35] Yet for me to know via the very reasoning with which I discern the transcendental gap between data and things the being that manifests this reasoning—this knowing might be weirdly like a serpent in a loop, swallowing its own tail. It is a profound paradox that what appears to be the nearest—my existence qua this actual entity, the shorthand for which is *human*—is phenomenologically the most distant thing in the universe. The supermassive black hole located at Sagittarius A in the center of the Milky Way, is far closer to my thought than my being human. The Muppets and their ilk actually *inhibit* the necessary ecological thought: the uncanny realization that every time I turned my car ignition key I was contributing to global warming and yet was performing actions that were statistically meaningless. When I think myself as a member of the human species, I lose the visible, tactile "little me"; yet it wasn't tortoises that caused global warming.

Fourthly, some of us are anxious that *Anthropocene* is hubristic, elevating the human species by assuming it has godlike powers to shape the planet. This is, on the face of it, infuriating—unfortunately not all humanists feel infuriated, trained as they are to suspect anything with "human" in it (in particular the Greek for *man*) and anything that seems like upstart straightforwardness, like using "we" in a lecture just because you think it might draw people together (wait a minute). But consider how it would sound as a rather eyebrow-raising defense. Say

I caused a car accident that killed your parents and your best friend. In court, I argue that it would be hubristic to blame myself. It wasn't really me, it was my right arm, it was the bad part of my personality, it was my car. Eyebrow-raising, and perfectly isomorphic with one mode of reactionary global warming denial: how dare we assume that much power over Nature! Now imagine that I represent the human species in a court in which many lifeforms are deciding who caused global warming. Imagine the "hubris" defense: "It would be hubristic of me to take full responsibility—after all, it's mostly the fault of this bad aspect of me, it was just an accident, I wouldn't have done it if I'd been riding a bike rather then using an engine . . . " Saying that the analogy doesn't work because I'm an individual just means you still have trouble, like most of us, thinking the concept *species*—which is the real problem.

The fact that humans really have become a geophysical force on a planetary scale doesn't seem to prevent the anxious spirits from accusing the term of hubris. Quibbling over terminology is a sad symptom of the extremes to which correlationism has been taken. Upwardly reducing things to effects of history or discourse or whatever has resulted in a fixation on labels, so that using *Anthropocene* means you haven't done the right kind of reducing. But what if you are not in the upward reduction business? Scientists would be perfectly happy to call the era Eustacia or Ramen, as long as we agreed it meant humans became a geological force on a planetary scale. Don't like the word *Anthropocene*? Fine. Don't like the idea that humans are a geophysical force? Not so fine. But the two are confused in critiques of "the anthropos of the Anthropocene." Consider that the term deploys the concept *species* as something unconscious, never totally explicit. No one decided in 1790 to wreck the planet by emitting carbon dioxide and related gases. Moreover, what is called human is more like a clump or assemblage of things that are not strictly humans—without human DNA for instance—and things that are—things that do have human DNA. Humans did it, not jellyfish and not computers. But humans did it with the aid of beings that they treated as prostheses:

nonhumans such as engines, factories, cows, and computers—let alone viral ideas about agricultural logistics living rent-free in minds. The reduction of lifeforms to prosthesis and the machination of agricultural logistics *is* hubristic, and tragedy (from which the term *hubris* derives) is at least the initial mode of ecological awareness. But this doesn't mean we are *arrogant* to think so.

Anthropocene is about humans—a mess of lungs and bacterial microbiomes and nonhuman ancestors and so on—along with their agents such as cows and factories and thoughts, agents that can't be reduced to their merely human use or exchange value. This irreducibility is why these assemblages can violently disrupt both use and exchange value in unanticipated (unconscious) ways: one cannot eat a Californian lemon in a drought. Returning to the point about intentions and hubris, "we" did it *unconsciously.* Becoming a geophysical force on a planetary scale means that no matter what you think about it, no matter whether you are aware of it or not, there you are, *being* that. This distinction is lost on some of those who react against the term. One cannot be hubristic about one's heartbeat or autonomic nervous system.

The fact that it is far from hubristic is also why geoengineers are incorrect if they think *Anthropocene* means we now have carte blanche to put gigantic mirrors in space or flood the ocean with iron filings. The argument for geoengineering goes like this: "We have always been terraforming, so let's do it consciously from now on."[36] Making something conscious doesn't mean it becomes nice. We have always been murdering people. How is deliberate murder more moral? Psychopaths are exquisitely aware of the suffering they consciously inflict. In relation to lifeforms and Earth systems, humans have often played the role of the Walrus concerning the oysters:

> "I weep for you," the Walrus said:
>   "I deeply sympathize."
> With sobs and tears he sorted out
>   Those of the largest size,

Holding his pocket-handkerchief
Before his streaming eyes.[37]

Consider the Freudian-slippy absurdity of James Lovelock's analogy of Jekyll and Hyde for science and engineering. Lovelock calls us the "species equivalent" of Robert Louis Stevenson's characters. It would only be faintly parodic to paraphrase his argument thus: "Only big science can save us. We know big science acted like Mr. Hyde for the last two centuries, but please know, we have a kindly inner doctor Jekyll. Let us be Jekyll. Please. Please trust us, *trust us*."[38] Unaware of its tone, Lovelock's argument sounds exactly like Mr. Hyde, as does Jekyll's own self-justification in the novel.

Unless we think the concept species differently, which is to say think humankind as a planetary totality without the soppy and oppressive universalism and difference erasure that usually implies, we will have ceded an entire scale—the scale of the biosphere, no less—to truly hubristic technocracy, whose "Just let us try this" rhetoric masks the fact that when you "try" something at a general enough level of a system, you are not *trying* but *doing* and changing, for good.

In any case one can't get rid of the unconscious that easily. Here is a sentence analogous to "We have always been terraforming, so let's do it consciously now": "I know I'm an addict so now I'm going to drink fully aware of that fact." Being aware of "unconscious biases" is a contradiction in terms. And there is a still more salient ecological observation we can make about the unconscious. Ecology, after all, is the thinking of beings on a number of different scales, none of which has priority over the other. When scaled up to what Douglas Kahn happily calls *Earth magnitude*, my conscious actions have an *unconscious* result that I did not intend.[39] Even when I am fully aware of what I am doing, myself as a member of the human species is doing something I am not intending at all and couldn't accomplish solo even if I wished it.

Humans created the Anthropocene—humans devised modes of agriculture we glimpse in Thomas Hardy's *Tess of the D'Urbervilles* that now cover most of Earth and are responsible for an alarming amount of global warming emissions all by themselves, let alone the carbon-emitting industry that agricultural mode necessitated. Not bacteria, not lemons. Such a making had unintentional or unconscious dimensions. No one likes having their unconscious pointed out, and ecological awareness is all about having it pointed out. As if in a disturbingly literal proof of Freud's refutation of the idea that the unconscious is a region "below" or "within" consciousness, we find the unconscious style of a certain mode of human being sprayed all over what lies outside the human, the biosphere. This unconscious is decidedly *(geo) physical*. The hint that there is an outside untouched by our conscious or explicit statements about what or how we dispose ourselves intellectually or culturally has become shocking or even taboo to some humanities scholars, right at the very moment when it would be handy if we could all be putting some effort into thinking this outside.

There are some substitutes for the term *Anthropocene*. For instance, I have been advised to call it *Homogenocene*. But this is just a euphemism. *Homogenocene* is true: humans have stamped their impression on things they consider as ductile as wax, even if those things cry. Yet, in a more urgent sense, the concept is false and anthropocentric. The iron deposits in Earth's crust made by bacteria are also homogeneous. Oxygen, caused by an unintended consequence of bacterial respiration, is a homogeneous part of the air. Humans are not the only homogenizers. Likewise, Haraway's and Latour's suggestion that we call it the *Capitalocene* misses the mark.[40] Capital and capitalism are symptoms of the problem, not its direct causes. If the cause were capitalism, then Soviet and Chinese carbon emissions would have added nothing to global warming. Even the champions of distributed agency balk at calling a distributed spade a distributed spade.

The concept of species, upgraded from the absurd teleological and metaphysical versions of old, isn't anthropocentric at all. Because it is via this concept, which is open, porous, flickering, distant from what is given to my perception, that the human is decisively deracinated from its pampered, ostensibly privileged place set apart from all other beings.[41]

"Anthropocene" is the first fully antianthropocentric concept.

Species at Earth Magnitude. When we scale up to Earth magnitude very interesting things happen to thinking. Some regularly suppose ecological statements to be universalistic generalizations: in large part they are adherents of capitalist economics, which finds the nonhuman structurally impossible to think, or Marxism, which doesn't find the nonhuman impossible to think—but which has imposed a host of inhibiting blocks to thinking the nonhuman. But thought at Earth magnitude isn't universalistic; it is highly accurate and specific. It is also deeply paradoxical in a way that reveals something basic to the structure of thought: a loop form.

I take Earth magnitude to mean "at a scale sufficient to open the concept *Earth* to full amplitude." Solar winds do this as they interact with Earth's magnetic shield and produce auroras. Global climate does this: the mass of terrestrial weather events are downwardly caused by a massive entity that exists at Earth magnitude. Human thought at Earth magnitude is human thinking that is as "large" as the aurora. It can think the aurora in such a way that its vastness is witnessed and opened in us. A single person can do this on the ground. You don't need to be a geostationary satellite or a scientist or an astronaut. Or a member of the UN or CEO of a global corporation.

We can now think species not as a thing we can point to, but as something like the aurora, a mysterious yet distinct, sparkling entity. It seems so easy: look, I'm a human, I'm not a duck or a doughnut. But this facile sense of ease is blocking something stupendously difficult: to follow and witness the being owing to which thinking is happening. Thinking goes into a loop. And the loop could be endless or not—we don't know yet and we might be pushing against the limits of computability if we try to know whether we will be looping forever. The thinking becomes a weird openness rather than cataloging and classifying, because it cannot presuppose a preformatted being as its content.

The Anthropocene is an antianthropocentric concept because it enables us to think the human species not as an ontically given thing I can point to, but as a hyperobject that is real yet inaccessible.[42] Computational power has enabled us to think and visualize things that are ungraspable by our senses or by our quotidian experience. We live on more timescales than we can grasp. Naomi Klein's description of global warming is good for hyperobjects in general: "Climate change is slow, and we are fast. When you are racing through a rural landscape on a bullet train, it looks as if everything you are passing is standing still: people, tractors, cars on country roads. They aren't, of course. They are moving, but at a speed so slow compared with the train that they appear static."[43] We are faced with the task of thinking at temporal and spatial scales that are unfamiliar, even monstrously gigantic. Perhaps this is why we imagine such horrors as nuclear radiation in mythological terms. Take Godzilla, who appears to have grown as awareness of hyperobjects such as global warming has taken hold. Having started at a relatively huge 50 meters, by 2014 he had reached a whopping 150 meters tall.[44] Earth magnitude is bigger than we thought, even if we have seen the NASA Earthrise photos, which now look like charming and simplistic relics of an age in which human hubris was still mostly unnoticed—relics of, precisely, a "space age" that evaporates in the age of giant nonhuman places. We have gone from having "the whole world in our hands" and "I'd like to buy the world a Coke" to realizing that the whole world, including "little" us, is in the vicelike death grip of a gigantic entity—ourselves as the human species. This uncanny sense of existing on more than one scale at once has nothing to do with the pathos of cradling a beautiful blue ball in the void.

Charles Long's 2014 *Catalin* installation at The Contemporary, an art museum in Austin, Texas, derived from the idea of hyperobjects some pieces Long calls *databergs*, impossible iceberglike chunks of absurdly disparate data: fatal car crashes in California versus fatal car crashes in Texas versus sea level rise observations and projections versus the U.S. unemployment rate for people over sixteen years old; fatal North American bear attacks versus Lamborghinis sold per year versus the percentage of jobs posted with *ninja* in the description or as an attribute versus quarterly global iPhone sales. . . . Such dizzying, hilarious icebergs of data are thinkable because hyperobjects are thinkable, hyperobjects that are melting actual icebergs.[45]

Humanistic tools for thought at Earth magnitude are lacking, and often because we have deliberately resisted fashioning them. For instance, dominant academic modes of cultural Marxism are hobbled by anthropocentrism. Such an anthropocentrism does indeed pick up on a strand in Marx's thinking in which the worst of architects is always superior to the best of bees. It is true that Marx himself gladly wrote about things outside the human sphere and outside the sphere of capital. However, the anthropocentric strain of cultural Marxism drastically foreshortens the nonhuman, casting nonhuman beings as mere aspects of human metabolic pathways. What such a Marxism calls *nature* is not actual trees and Arctic foxes but trees and foxes as they are metabolized by human economic relations. Use value isn't "what things really are for," but "what things are for humans." In this sense even Aristotelian definitions of things via their final cause are more embracing.

When Marx talks about the depletion of the soil, he isn't worrying about earthworms and bacteria. Marx is concerned about the human capacity to metabolize enough energy to remain in existence.[46] But even the soil, in this narrow correlationist sense, is a bit too dirty for some forms of cultural Marxism to mention.[47] This correlationist anxiety about the real within Marxisms emerges simultaneously with the creeping awareness that factoring *energy* throughput (oil, solar, natural gas, wind, coal . . .) into historical accounts of social space necessarily and scandalously generates a bigger picture than the one provided by the notion that human economic relations and the class struggle they entail are what make things real: "All narratives about the changes in the human condition are narratives about the changing exploitation of energy sources—or descriptions of metabolic regimes. This claim is not only one dimension more general than the Marx-Engels dogma that all history is the history of class struggles; it is also far closer to the empirical findings. Its generality extends further because it encompasses both natural and human energies."[48] "One dimension more general": Sloterdijk's telling phrase says it all. This is about scale and how humans now find themselves outscaled, caught in and concerned for all kinds of nonhuman *place*. Place is no longer simply human. A huge swath of terrestrial reality is unaccounted for in traditional Marxism. That's what happens when, like Kant, one restricts the decision as to what counts as real to one corner of the universe: in Kant's case, the gap between the (human) subject and everything else; in Marx's case, the gap between (human) economic relations and everything else.

It might be argued that "livestock" are as much the proletariat as human workers.[49] The etymology that associates patriarchal property (chattel) with nonhumans (cattle) with standing reserve (capital) makes this quite obvious.[50] It might be the case that, for the specter of communism to haunt earth sufficiently, the specter of the nonhuman would need to be embraced by the specter of communism. Full communism might need to include earthworms and bacteria, although for reasons given in the Third Thread that might look more like anarchic clusters than one system to rule them all. How can we think totality and collectivity at a moment when there is no good reason to stop at a certain species or scalar boundary? For this is what we should task ourselves with: thinking future coexistence, namely coexistence unconstrained by present concepts.

The best of bees. Marx writes that the best of bees is always worse than the worst of architects.[51] That's because the architect is imagining her or his building and the bee is just executing an algorithm. We could go about disproving the claim in two ways: (1) considering the bees and (2) considering ourselves. Let's examine both in turn.

(1) We could set up a lot of expensive experiments to find out whether bees imagined things. Of course we would have to know what we were looking for, namely empirical evidence of imagination. For instance, we could find out whether bees hesitated. If they hesitated or looked around while they were carrying out a task, that might be evidence that they weren't just blindly following an algorithm.

So defensive can some Marxists become concerning this point in Marx—they do perhaps sense the danger—that they sometimes assert this passage is just metaphorical. That is to say some Marxists claim that by *bees* Marx really means workers and by *architects* Marx really means the bourgeoisie. Yet, if anything, that is more insulting still, and not only to bees. According to this interpretation Marx is saying that workers just blindly execute. How on earth are these poor crude androids going to figure out what's going on and start a revolution? And how could they ever fulfill human species-being, the Marxist concept that pictures humans imaginatively creating their own environments? The workers would have to leave species-being fulfillment to the architect, and even a sloppy one would do a better job than them.

It has indeed been shown that ants climbing up little ladders look around them rather than walking up automatically. They weigh options when it comes to where to live and so on.[52] Such findings suggest that ants anticipate and assess situations, which is at least part of what architects are supposed to do when they design a building. It has also been shown that bees build mental maps to find their way home—they aren't just on autopilot.[53] We are beginning to allow that nonhumans have minds. Creative experiments have shown that rats experience regret.[54] The problem with disproof tactic (1), however, is that our poor scientist has to know roughly what she is looking for already before running the experiment, and this means that she is forever haunted by a deep problem that affects both science and humanities in the Anthropocene, the age of Hume: the age in which there is no objectified, obvious cause and effect churning away below phenomena like cogwheels. Cause and effect are inferences we make concerning statistical correlations in data. (Incidentally, accurate correlations in *ecological* data, since ecological reality is so rich and ambiguous, are notoriously difficult to find.)[55]

Cause and effect are "in front" of things, not behind them: in front ontologically rather than spatially.[56] Which is to say that in order for there to be causality there must always already be objects. In this sense, weird as it is to say so given our tendency to snap back to mechanistic causal theories, causality in a post-Newtonian world has its rightful place *in the aesthetic dimension*.

Scientists are now beginning to figure out something we've known in the humanities and arts for some time: one is entangled with the data one is studying. Kant grounded Hume's insight about causality in just this thought, which we now call correlationism. We can't see things in themselves, we can see human-flavored correlates of those things. But there *are* things in themselves. So we are caught in a dilemma, whose name is *hermeneutic circle*. Scientists call it *confirmation bias*, which is why only a small percentage of physicists now think that physics is saying anything true at all about reality.[57] They are justifiably concerned by a basic implication of Hume that *scientism*, not science, has been blocking for two hundred years. Since some of us are scientistic even if we are scientists, this isn't surprising, scientism being in a way a method of shutting one's ears to what is most interesting about science as such. Science swears off making ontological statements of any kind, an abstention that makes you a scientist far more than the Hippocratic oath makes you a doctor.

The term *confirmation bias* is itself a symptom of some kind of confirmation bias . . . "Confirmation bias" suggests that there are things over there and interpretations over here, and that those interpretations can therefore be biased. But this idea of objects over there and subjects over here is precisely what correlationism and its consequent hermeneutic circle are saying is illegal—it's a metaphysical factoid that you've smuggled into your view pretheoretically. Never mind that Kant himself had smuggled in this view, which is the old Aristotelian—and I shall argue *agricultural*—picture of bland substances decorated with accidents.[58] That's exactly what we *can't* assume things are like. It's the kind of thing that gives rise to ideas that bees are just blind robots while architects are gravity-defying subjects. Heaven help us, we would never ever want to be denigrated to the status of a thing, because we all "know" in advance that things are lumps.

The prejudice that things are lumps is one reason why object-oriented ontology has come in for criticism. By saying *object,* OOO touches a third rail. Within that there is an even more sensitive third rail of beliefs about what entities are, sensitive because of its political implications, sensitive furthermore because those beliefs were hardwired into Earth's surface in a way so effective that millions of lifeforms are now going extinct. In 2014 the World Wildlife Fund revealed that 50 percent of animals (lifeforms in the animal kingdom) had disappeared in the last four decades.[59] Noticing that fact is horribly uncanny: we want to go on dreaming our anthropocentric dream because it feels safer. Despite its provocative use of the word *object,* OOO is the diametrical opposite of the dream. OOO might be a mode of waking up.

Now let us consider the second disproof tactic.

(2) The lack of obvious empirical evidence concerning imagination points to a much more efficient and much cheaper way of proving whether or not the best of bees is always worse than the worst of architects. What do we have already? We already have some sense that bees and ants can do things that look like things that we can do with our minds. So by inference we aren't as special as we thought. But we can take a step back and think about the really obvious state of affairs, which is that *we lack reliable empirical evidence for imagination as such*. I'm not saying there is no imagination. Far from it. What

I require the Marxist to do is to prove that the *architect* has imagination. Prove that I have imagination, as a human being. Prove that I'm not executing an algorithm. More to the point, prove that my idea that I'm not executing an algorithm isn't just the variety of algorithm that I've been programmed to execute.

As we'll often see in *Dark Ecology*, being paranoid that I might *not* be a person is in fact a default condition of *being* a person. There is a profound philosophical hesitation here. Because it's so stimulating, we usually like to collapse the duality into one of its terms. We could decide that there is no imagination, that we are totally conditioned, a thought that is usually close to reducing things to matter. Thoughts are functions of brains or something, perhaps in the strong "eliminative materialist" sense: if we can explain mind in terms of brain there is *no mind at all*: the mind is a pure illusion. The mind, on this view, isn't even an emergent property of a brain. Or we could go the other way and say that there is personhood and that it's totally different from being a determined machine. We could perhaps back this up with some idea of mental qualia or the irreducibility of consciousness. What's interesting is that we are trying to get rid of a profound wonderment. And since, along with Plato, I take wonderment to be the basic phenomenological chemical of philosophy, we are implicitly trying to shut down philosophy when we take these paths.

If you have some hesitation or difficulty proving that humans imagine, that's fantastic. It means that you have accepted modern science, which means you have accepted modern philosophy since the start of the Anthropocene. And if you try not to collapse the hesitation, like the hesitation of an ant on a tiny ladder, that's even better. It means you have accepted the deep reason for the validity of modern science and philosophy. You have not collapsed the wonderment. You have become scientific, but not scientistic. You are refusing to pounce on things with metaphysics. You are beginning the difficult upgrade of concepts such as *person* and *thing* and *species* so essential to human thought in an ecological age, and indeed so essential for the continued existence of lifeforms.

You are beginning to think at Earth magnitude.

At Earth magnitude, anthropocentric distinctions don't matter anymore. Or, better, they cease to be thin and rigid. They matter amazingly differently. These distinctions include binaries such as *here* versus *there*, *person* versus *thing*, *individual* versus *group*, *conscious* versus *unconscious*, *sentient* versus *nonsentient*, *life* versus *nonlife*, *part* versus *whole*, and even *existing* versus *nonexistence*. Biology raises the problem of life as such. That the life-nonlife boundary isn't exactly erased, instead becoming far less thin and rigid, is an issue within biology as it begins to go into crisis, insofar as this boundary is found to be more than trivially flexible. Some, for instance, are wondering whether evolution is restricted only to organic chemicals. At the boundary between biology and chemistry, Darwin is of surprisingly little use unless we boldly extend his insights to include something like natural selection at the chemical level.[60] As we shall see, this is about how fundamental pattern making is to reality, because patterns are the basis for replication.

The same upgrade happens to sentience, consciousness, and, in an ecological age, between the human and the nonhuman altogether, notably such that ideas like *world* and *here* begin to look not like big abstract concepts but rather small, localized, human flavored. Let us reiterate: this is not because there is no such thing as place. As I observed earlier, in evolution science you can't look at a duck and see what it's "for" in some obviously human-flavored way. Ducks aren't *for* anything. Teleology has evaporated, hierarchies have collapsed; but there are still ducks and humans and Earth, and sentience and lifeforms as opposed to salt crystals. They become more and more vivid as our ways of distinguishing them become more and more questionable.

This may not have been what we were expecting. We might have been expecting that, on a much larger scale, things would become much easier to understand. Indeed, we might criticize those who tried to think at larger scales for being simplistic. We might even argue that they were deluded. We might accuse someone of being a bit of a hippie for talking at scales beyond the human. We think that the hippie is ideologically deluded into saying things can matter (become "real") outside human economic mediation. All that *we are the world* and *save Earth* stuff is bourgeois pabulum meant to keep us docile.

Our Marxist has this allergic reaction because he or she is rigidly adhering to a solution to the Kantian shock—the shock that there are things, but that when we look for them we find only human-flavored thing data. We never see the actual raindrop; we have raindrop feelings, raindrop thoughts, raindrop perceptions.[61] Kant himself tries to contain the explosion by saying that there is a top-level way of understanding the raindrop, namely mathematizing it via a concept of extension as the bedrock of what a thing is. The transcendental subject is the being that decides whether a thing is real or not. Post-Kantians contain the explosion two ways. Either they reduce everything to matter and ignore the implications of modern philosophy and the science derived from it. Or they wish away the gap between phenomenon and thing by claiming more strongly than Kant that some kind of Decider makes the thing real. A succession of hopeful substitutes for the Kantian subject arises: *Geist* (Hegel), will (Schopenhauer), will to power (Nietzsche), Dasein (Heidegger).

And, in the case of Marx, *human economic relations* make things real. And, in the hardcore Hegelian Lacanian Althusserian version, these relations are an *in-the-last-instance* that determine everything else like the sucker of a giant and sprawling undersea creature, attached to a rock in one place, but attached really strongly, incapable of being peeled off that rock. So that for the cultural Marxist, unconsciously retweeting a substance-and-accidents model of things, there is ideology (accidents) and human economic relations (substance).

By putting it this way, I have already committed a horrible sin because I have used the word *human*. By using that word I have implied that there might be a world or worlds beyond or different than the human, which is as good as saying that there are such worlds. I have broken a taboo in implying Marxism doesn't explain everything, because there are cats, coral, and galaxies. The very concept *ecology*, coined by Ernst Haeckel, was a way to say *the economy of nature* in a compact way. Beavers and spiders and bacteria metabolize things too. Species-being isn't what it's cracked up to be.

This humiliatingly means that, claims to the geopolitical notwithstanding, cultural Marxism cannot think the *geo* sufficiently to think the geopolitical. So cultural Marxism lets fly a volley of accusations against the sinners: they are racists or sexists or colonialists because they use concepts such as species. Either you are into feminism or you are a speculative realist.[62] The same brittle theistic logic was deployed by the Bush administration with its "you are either with us or with the terrorists."

Backed into a corner and reduced to apoplectic double binds, the accusers conceal a genuine anxiety: species in the nonteleological sense is what Marxism cannot think. Despite Marx's having written a fan letter to Darwin, the Marxist notion of species-being still adheres to teleology in the sense that, according to the extreme correlationist definition, humans are "for" creating their own environment, and this is unique—just try to forget about ants and beavers. The inability to think species is despite Marx's grounding in Feuerbach, whose whole project was to show that species was not at all an abstract, universalist generalization but a finite, concrete entity, albeit one that exists at a scale larger than the one on which we normally think. Species-being fits in the lineage of Aristotle. Humans produce, which means they imagine, unlike bees, which (I suspect which, rather than who, for users of this concept) are just robots. And robots are just things. And things are inanimate, unconscious, lumps of whatever decorated with accidents.

Let's remind ourselves right now that this problem applies in thick spades to capitalist economic theory too. Capitalist economics is also an anthropocentric practice that has no easy way to factor in the very things that ecological thought and politics require: nonhuman beings and unfamiliar timescales. Considering public policy at timescales sufficient to include global warming, economic theory tends to throw up its hands and say, "This doesn't fit our science"—well duh.[63] What is really meant here is "This doesn't effect our interpretation of data given that, unlike a physicist, we are unwilling to notice that we may suffer from confirmation bias." Or consider the argument within economics that depression about ecological issues is dangerous or absurd or impossible—how it can be all three without being a politicized pseudotarget eludes me, but the idea is again that "the science" doesn't justify it: why on Earth would anyone want to impose a tax on goods entering or leaving the country unless one were some kind of "authoritarian" hostile to "free trade?"[64] Such reasoning is deaf to the nonhumans whose inclusion in thought compels one to think about, for example, minimizing or changing one's energy use, perhaps by taxing things that have to travel a long way. Psychology and economics, "sciences" closest to humans, are, not surprisingly, deeply anthropocentric and unwilling to consider that they may be caught in hermeneutical loops.

Thinking the human at Earth magnitude is utterly uncanny: strangely familiar and familiarly strange. It is as if I realize that I am a zombie—or, better, that I'm a component of a zombie despite my will. Again, every time I start my car I'm not meaning personally to destroy lifeforms—which is what "destroying Earth" actually means. Nor does my action have any statistical meaning whatsoever. And yet, mysteriously and disturbingly, scaled up to Earth magnitude so that there are billions of hands that are turning billions of ignitions in billions of starting engines every few minutes, the Sixth Mass Extinction Event is precisely what is being caused. And some members of the zombie have been aware that there is a problem with human carbon emissions for at least sixty years. The first global warming evidence was published in 1955.[65] Humans have now ensured over 400 parts per million of carbon dioxide in Earth's atmosphere. Arctic temperatures are at the highest they have been for 44,000 years.[66]

It doesn't seem to matter whether I'm thinking about extinction or not, whether I mean to or not, even whether or not I start my own personal car! So, back to that question: am I conscious? Prove that I'm not better than the best of bees. Prove that my idea of consciousness, let alone individual free will, isn't just the algorithm that my particular species has evolved to run. Stripped of its metaphysical, easy-to-identify, soothingly teleological content, the notion of species is an uncanny thought happening not in some universal or infinite realm but at Earth magnitude. It is strictly uncanny in the Freudian sense, if we recall that Freud argues that uncanny feelings in the end involve the repressed intimacy of the mother's body, the uterus and the vagina out of which you came.[67] This is significant because thinking this mother's body at Earth magnitude means thinking ecological embodiment and interdependence. That uterus is not just a symbol of the biosphere, nor even an indexical sign of the biosphere, pointing to it like a footprint or a car indicator. The uterus *is* the biosphere in one of its manifold forms, just as me turning the key in the ignition *is* the human in one of its manifold forms. It is, and it isn't, which is how you can tell it's real. To be real is to be contradictory, to be a member of a set that doesn't include you. To be real is not to be easy to identify, easy to think, metaphysically constantly present.

When we think species this way, we see global warming as a *wicked problem*—or even as a *super wicked problem*.[68] A wicked problem is one you can rationally diagnose but to which there is no feasible rational solution. There are four main aspects:

(1) Wicked problems are unique and thus *irreducible* and difficult to conceptualize and anticipate. Likewise, they are unverifiable. If we "solve" global warming, we will never be able to prove that it would have destroyed Earth . . .

(2) Wicked problems are *uncertainly interminable*: there is no way to predict when the problem will have ceased to function.

(3) Wicked problems are *alogical* in the sense that solutions to them cannot be assessed as right or wrong, but rather as good or bad. There is a sharp division between ethics and ontology here, one that we think we like ("You can't get an *ought* from an *is*"), but that in practice we hate: we contemporary humanists usually want ideas about reality bundled with an easy to identify politics.

(4) *Irreversibility*—there are no trial runs, no reverse gears, no attempts to solve wicked problems, only actual solutions that drastically alter things.

There appears to be no way to solve a wicked problem neatly and know that we have solved it. Like poems, wicked problems entangle us in loops. We know that our reading of a poem is provisional and that our thoughts about what poems are influence how we read them; the same goes for global warming. Wicked problems make the strange loop form of ecological beings obvious. As a matter of fact, global warming is a "super wicked problem": a wicked problem for which time is running out, for which there is no central authority; those seeking the solution are also creating it, and policies discount the future irrationally.[69] The superness has to do with how we are physically caught "in" the problem: the damaged biosphere. We are thus in an obvious looplike relationship with the problem. In a weird loopy not-quite inversion of the song, *the whole world has got us in its hands*—because we became a geophysical force.

Wicked problems have uncertain boundaries because they are always symptoms of other problems. Global warming is a symptom of industrialization, and industrialization is a symptom of massively accelerated agriculture. Of what is this acceleration a symptom? We could say that it was capitalism, but that would be circular: accelerating agriculture and subsequent industrialization are symptoms of capitalism, not to mention existing forms of communism. So we are looking for the problem of which these things are symptoms. What is it? Why, if so influential, is it so hard to point to?

Agrilogistics. Two reasons: it is everywhere and it is taboo to mention it. You could be labeled a primitivist even for bringing it up.

> In the Golden Age, agriculture was an abomination. In the Silver Age, impiety appeared in the form of agriculture. In the Golden Age, people lived on fruits and roots that were obtained without any labor. For the existence of sin in the form of cultivation, the lifespan of people became shortened.[70]

> I have placed a curse on the ground. All your life you will struggle to scratch a living from it. It will grow thorns and thistles for you, though you will eat of its grains. All your life you will sweat to produce food, until your dying day. Then you will return to the ground from which you came. For you were made from dust, and to the dust you will return.[71]

Two ancient texts written within agricultural temporality condemn agriculture, and rather startlingly accurately: the science is on their side.[72] Consider the collection *Paleopathology at the Origins of Agriculture*. The very title fleetingly suggests that there was *an ancient pathology* (paleopathology), as if the origins of agriculture were pathological. It is as if science couldn't help employing the rhetoric of agricultural religion, as if science itself were suspended in agricultural time. This rhetoric pits agriculture against agriculture in what we could call agricultural *autoimmunity*, an agricultural allergy to itself. Foundational Axial (agricultural) Age stories narrate the origin of religion as the beginning of agricultural time: *an origin in sin*. The texts are almost shockingly explicit, so it's strange we don't think to read them that way. Pretty much out loud, they say that religion as such (was there "religion" beforehand?) was founded in and as *impiety*. And the thistles keep growing, the sweat keeps pouring, and humans are from dust, not from themselves as later agricultural myths (from the Theban cycle to the Enlightenment) will proclaim. We witness the extraordinary spectacle of "religion" talking about itself as a reflective, reflexive loop of sin and salvation, with escalating positive feedback loops. Like agriculture.

Now consider this text. The author is looking down on a valley in China: "Forest—field—plow—desert—that is the cycle of the hills under most plow agricultures. . . . We Americans, though new upon our land, are destroying soil by field wash faster than any people that ever lived. . . . We have the machines to help us to destroy as well as to create."[73] It is 1929. Apart from noting the time span between these three texts, need one say more?

What is this "human" species, instigator of the Anthropocene, fragile sand drawing? Evidently the term as used here is not essentialist, if *essentialist* means believing that how things exist is that they are constantly, metaphysically present. This is the very metaphysics that isn't strictly thinkable in the lineage of Kant and his subsequent lineage holders, including Heidegger, who inspired Lacan, who taught Foucault, who told us of human faces drawn in the sand. Not thinkable, that is, if you want to be modern—and not thinkable in the sense that unsustainable paradoxes arise when you try to think this way.

Beliefs in constant presence derive ultimately from a default ontology persistent in the long moment in which the Anthropocene is a disturbing fluctuation. We are still within this twelve-thousand-year "present" moment, a scintilla of geological time. What happened in Mesopotamia happens "now," which is why it has made sense for *Dark Ecology* to refer to us as Mesopotamians. This long now started somewhere, sometime. It is bounded. Yet to think outside it, since that very outside is defined by it, is to think within it. The contemporary phenomenon of the gluten-free diet perfectly embodies this—the diet that currently obsesses almost a third of American adults, despite actual gluten intolerance affecting a single-digit percentage of the population. The gluten-free phenomenon is a magic bullet solution to modernity. "Gluten" sounds abject and glutinous, and it is found in wheat: agriculture as sin, just like Genesis says. Such an allergic reaction of modernity to itself is absurd: even Neanderthals made bread. Consider the modern hatred of the body that links with a profound (and accurate) unease that "something is wrong" and is then blended with primitivism: the "Paleo diet." The term *Paleo* acknowledges that something is wrong with the Neolithic, the term we use for post-Mesopotamian human social forms: something is wrong—*as Genesis had already pointed out.*

Remember Earth clearly. Thinking outside the Neolithic box would involve seeing and talking at a magnitude we humanists find embarrassing or ridiculous or politically suspect. Perhaps it is completely outlandish: thinking this way is easily marginalized as an activity for loons. We can find examples, but they are indeed marginal. We might for instance find them in the insights of psychedelic drug-fueled depression exemplified in the middle-period work of the British techno group Orbital (active since 1989). The video for Orbital's "The Box" is a miracle of juxtaposed timescales. A lonely wanderer played by Tilda Swinton holds a position for a very long time. A camera films her and what happens around her. Then the film is sped up, so that the wanderer appears to be walking through a megacity while cars and people rush around her at breakneck speed. The physical difficulty of the dancer's role is breathtaking, which performs the difficulty of thinking on more than one scale at once: the thinking that ecological awareness demands.

The dancer stops outside a cheap electronics store. She watches televisions in the window. Unbeknownst to the passersby, since it is happening on such a slow timescale relative to them, secret messages are flashing on the screen. Only the isolated wanderer can see them:

she functions in a temporal scope sufficient to read the messages that perhaps to others appear only as minuscule flickers. One has to pause the video to read the evocative sequence oneself:

REMEMBER EARTH CLEARLY
BAD
DAMAGED
BATTERED
PLANET
FRAYED
DUSTBOWL
COMPROMISED
WAKE UP
MONSTERS EXIST[74]

It's a sinister, paranoid moment of ecological awareness. What is the monster? Sophocles encapsulated it already in the astonishing Second Chorus of his Theban play *Antigone*: *Of the many disturbing beings, man is the most disturbing*. Why? Because he plows, and because he is aware of how this plowing disturbs Earth. "DUSTBOWL" obviously references the disaster of agricultural feedback loops. We are disturbed by our disturbance—and we don't stop: seeing "MONSTERS EXIST" on a TV screen in a shop window is like the fantasy of seeing a monstrous face in the mirror when you pass by in the dark.

Imagine *seeing* on more than one timescale—just as geology and climate science *think* on more than one. Imagine for a moment that the phenomenon-thing gap were closed and that you could see everything. This is what is happening to the woman in *The Box*. The lonely walker perceives the phrase "MONSTERS EXIST" on a television screen that no one else can see: they would require the scaled-up temporality at which she is living to see it. This is like being able to see hyperobjects. Why is this disturbing? Because *you are already living* on more than one timescale. Ecological awareness is disorienting precisely because of these multiple scales. We sense that there *are* monsters even if we can't see them directly.

There's a monster in the dark mirror, and you are a cone in one of its eyes. When you are sufficiently creeped out by the human species, you see something even bigger than the Anthropocene looming in the background, hiding in plain sight in the prose of Thomas Hardy, the piles of fruit in the supermarket, the gigantic parking lots, the suicide rate. What on Earth is this structure that looms even larger than the age of steam and oil? Isn't it enough that we have to deal with cars and drills? Hardy provides a widescreen way of seeing agricultural production, sufficient for glimpsing not only the immiseration of women in particular and the rural working class in general at a specific time and place but also the gigantic machinery of agriculture: not just specific machines, but *the machine that is agriculture as such*, a machine that predates Industrial Age machinery. Before the web of fate began to be woven on a power loom, machinery was already whirring away.

A brief history of agrilogistics. Dark Ecology is going to call this twelve-thousand-year machination *agrilogistics*. The term names a specific logistics of agriculture that arose in the Fertile Crescent and that is still plowing ahead. Logistics, because it is a technical, planned, and perfectly logical approach to built space. Logistics, because it proceeds without stepping back and rethinking the logic. A viral logistics, eventually requiring steam engines and industry to feed its proliferation.

Agrilogistics: an agricultural program so successful that it now dominates agricultural techniques planetwide. The program creates a hyperobject, global agriculture: the granddaddy hyperobject, the first one made by humans, and one that has sired many more. Toxic from the beginning to humans and other lifeforms, it operates blindly like a computer program. The homology is tight since algorithms are now instrumental in increasing the reach of agrilogistics. Big data makes bigger farms.[75]

Agrilogistics promises to eliminate fear, anxiety, and contradiction—social, physical, and ontological—by establishing thin rigid boundaries between human and nonhuman worlds and by reducing existence to sheer quantity. Though toxic, it has been wildly successful because the program is deeply compelling. Agrilogistics is the smoking gun behind the smoking chimneys responsible for the Sixth Mass Extinction Event. It isn't difficult to find a very brief example of the scope of agrilogistics in the fact that Europeans tolerate milk. A genetic mutation was encouraged to flourish within a few thousand years of original Fertile Crescent farmers, who had already reduced the lactose content in their cows' milk. Humans with this mutation became aggressive vectors for agrilogistics, and agrilogistics wiped out indigenous European human social forms.[76]

The humanistic analytical tools we currently possess are not capable of functioning at a scale appropriate to agrilogistics because they are themselves compromised products of agrilogistics. The nature-culture split we persist in using is the result of a nature-agriculture split (*colo, cultum* pertains to growing crops). This split is a product of agrilogistic subroutines, establishing the necessarily violent and arbitrary difference between itself and what it "conquers" or delimits. Differences aside, the confusions and endlessly granular distinctions arising therefrom remain well within agrilogistic conceptual space.[77]

Agrilogistics arose as follows. About 12,500 years ago, a climate shift experienced by hunter-gatherers as a catastrophe pushed humans to find a solution to their fear concerning where the next meal was coming from. It was the very end of an Ice Age, the tail end of a glacial period. A drought lasting more than a thousand years compelled humans to travel farther. It happened that in the Fertile Crescent of Mesopotamia barley and wheat were growing wild beneath the trees. The same can be said for rice growing in China; corn, squash, and beans growing in America; and sorghum and yam in Africa.

Significantly, the taro of New Guinea is hard to harvest and low in protein, not to mention hard to plant (you have to plant taro one by one), and so the farmers in the highlands never "advanced" from hunter-gathering. The taro cannot be *broadcast*. Incidentally, so many terms from agrilogistics have become terms in media (*field* among them), not to mention the development of that very significant medium, writing. How we write and what we write and what we think about writing can be found within agrilogistics.

Humans in Mesopotamia established villages with granaries. The storage and selection of grain pushed the harvested plants to evolve. Humans selected grain for its tastiness, ease of harvesting, and other criteria favored by the agrilogistic program. Scaled up, the evolutionary pressure was substantial. Nine thousand years ago humans began to domesticate animals to mitigate seasonal variations in game, a modification to the agrilogistic program that kept it in existence.[78] Several agrilogistic millennia later, domesticated animals far outweigh (literally) the nondomesticated ones. Humans represent roughly 32 percent of vertebrate biomass. The other 65 percent is creatures we keep to eat. Vertebrate wildlife counts for less than 3 percent.[79] The term *cattle* speaks to this immensity and to a too-easy ontology humming away in its background.

Miserable social conditions were the almost immediate consequence of the inception of agrilogistics, yet the virus persisted like an earworm or a chair, no matter how destructive to the humans who devised it.[80] Private property emerged, based on settled ownership and use of land, a certain house, and so on. This provided the nonhuman basis of the contemporary concept of self, no matter how much we want to think ourselves out of that. Agrilogistics led rapidly to patriarchy, the impoverishment of all but a very few, a massive and rigid social hierarchy, and feedback loops of human-nonhuman interaction such as epidemics.[81]

Despite the misery and disaster, agrilogistics continues to run. For all intents and purposes, agrilogistic boiling is performed *for its own sake*—there have been no other great reasons, as we shall see. That is very strange, because growing and nurturing theories of ethics based on self-interest is a major feature of agrilogistics. Yet, in practice, it is as if humans became fascinated with maintaining the program at whatever cost to themselves. The loop of agrilogistics for agrilogistics' sake should remind one of the aestheticism of "art for art's sake." It is an unorthodox aestheticism of utility, *an aestheticism of the nonaesthetic*. The non- or even antiaesthetic is intrinsic to agrilogistic production: humans evolved wheat, for instance, for minimal flowers and maximal nutrition. So-called utlility operates just like so-called inutility.

The idea that humans began "civilization" in Mesopotamia is a retroactive positing if ever there was one. Humans looked back and designated the time of early agrilogistics as a unit, justifying the present as if civilization had suddenly emerged like the goddess Athena from the head of the human without any support. Without coexistence. "Civilization" was a long-term collaboration between humans and wheat, humans and rock, humans and soil, not out of grand visions but out of something like desperation. When one includes the nonhumans previously imaged as "nature" so as to airbrush smooth the agrilogistical project, the story of civilization is even simpler: "We turned the region into a desert, and had to move west." The poems of Percy Shelley often speak of ancient patriarchal monotheist tyrants ruling deserts in Egypt or Persia, leaving behind a broken statue sneering in the sandy emptiness: "Nothing beside remains.... The lone and level sands stretch far away."[82] For *civilization,* read *agrilogistic retreat*.

The human hyperobject (the human as geophysical species) became a machine for the generation of hyperobjects. Precisely because of the sharp imbalance between the "civilization" concept and actually existing social space (which was never fully human), agrilogistics itself having produced this difference, "civilizations" (the human structures of agrilogistic retreat) are inherently fragile. Just as World War II was the viral code that broke the program of a certain imperialism, one wonders whether global warming will be the viral code that breaks the machinations of a certain neoliberal capitalism and whether this will shut down agrilogistics itself. One wonders. And what a price to have paid, in part because agrilogistics underlies all "civilized" forms thus far, from slave-owning societies to Soviets.

The very concept of "world" as the temporality region suffused with human destiny emerges from agrilogistic functioning. World, as Heidegger knew, is *normative*: the concept works if some beings have it and some don't. When, like Jakob von Uexküll, you start to realize that at least all lifeforms have a world, you have begun to cheapen the concept almost to worthlessness. The concept reaches zero when humans realize that there is no "away," that there is no background to their foreground despite the luxury holiday ads, a lack of a stage set on which *world* can perform, a lack that is evident in the return of culturally (and physically) repressed "pollution" and awareness of the consequences of human action on nonhumans. The end of the biosphere as we know it is also the end of the "world" as a normative and useful concept.

The three axioms of agrilogistics. We live inside a philosophy alongside worms, bees, plows, cats, and stagnant pools. But the philosophy is silent or, as Anne Carson might say, "terribly quiet"; it betrays itself in the movements of Tess in the field and in the form of the field itself, but agrilogistics is a dumb show so familiar that it's almost invisible: the silent functioning of metaphysics. One goal of *Dark Ecology* is to make agrilogistic space speak and so to imagine how we can make programs that speak differently, that would form the substructure of a logic of future coexistence.

The agrilogistic algorithm consists of numerous subroutines: eliminate contradiction and anomaly, establish boundaries between the human and the nonhuman, maximize existence over and above any quality of existing. Now that the logistics covers most of Earth's surface, even we vectors of agrilogistics, Mesopotamians by default, can see its effects as in a polymerase chain reaction: they are catastrophically successful, wiping out lifeforms with great efficiency.

Three philosophical axioms provide the logical structure of agrilogistics:

(1) The Law of Noncontradiction is inviolable.
(2) Existing means being constantly present.
(3) Existing is always better than any quality of existing.

We begin with axiom (1). There is no good reason for it. We shall see that there are plenty of ways to violate this law, otherwise we wouldn't need a rule. This means that axiom (1) is a prescriptive statement disguised as a descriptive one. Formulated rightly, axiom (1) states, *Thou shalt not violate the Law of Noncontradiction*. Axiom (1) works by excluding (undomesticated) lifeforms that aren't part of your agrilogistic project. These lifeforms are now defined as pests if they scuttle about or weeds if they appear to the human eye to be inanimate and static. Such categories are highly unstable and extremely difficult to manage.[83]

Axiom (1) also results in the persistent charm of *the Easy Think Substance*. Agrilogistic ontology, formalized by Aristotle about ten thousand years in, supposes a being to consist of a bland lump of whatever decorated with accidents. It's the Easy Think Substance because it resembles what comes out of an Easy Bake Oven, a children's toy. Some kind of brown featureless lump emerges, which one subsequently decorates with sprinkles.

In Tom Stoppard's play *Darkside*, which magically lets Pink Floyd's *The Dark Side of the Moon* speak its implicit ecological philosophical content, a cynical philosophy teacher explains the famous trolley problem. If there are lots of people on a train heading over a cliff, it is ethical to switch the points to divert the train, even if the train runs over a single person stuck on the track onto which the train diverts.

When a sensitive student asks the teacher about the experiment ("Who was on the train?" "Who was the boy?"), the teacher insists that it's merely a thought experiment, that there's no point in knowing. Yet this perceived irrelevancy is normative: it is what generates the utilitarianism in the first place.

The girl student, dismissed as insane, asks the teacher, "Who was on the train?" The teacher responds, "We don't know who was on the train, it's a thought experiment."[84] The humor compresses an insight: this nondescription of Easy Think passengers implies an unexamined thought that gives no heed to the qualities of the people on board. Only their number counts, *the fact that they merely exist*. Existing is better than any quality of existing, according to axiom (3). It doesn't even matter how many *more* people there are. Even the sheer quantity of existing is treated as a lump of whatever. Say there were three hundred people on the track and three hundred and one people in the train. The train should divert and run over the people on the track. More to the ecological point, imagine seven billion people on the train and a few thousand on the track. This represents the balance (or lack thereof) between the human species and a species about to go extinct because of human action. This amazing pudding of stuff isn't even a fully mathematizable world. Counting itself doesn't count. For a social form whose new technology (writing) was so preoccupied with sheer counting, as surviving Linear B texts demonstrate, this is ironic.

The lump ontology evoked in axiom (1) implies axiom (2): to exist is to be constantly present, or the *metaphysics of presence*. Correctly identified by deconstruction as inimical to thinking future coexistence, the metaphysics of presence is intimately bound up with the history of global warming. Here is the field: I can plough it, sow it with this or that or nothing, farm cattle, yet it remains constantly the same. The entire system is construed as constantly present, rigidly bounded, separated from nonhuman systems. This appearance of hard separation belies the obvious existence of beings who show up ironically to maintain it. Consider the cats and their helpful culling of rodents chewing at the corn.[85] The ambiguous status of cats is not quite the "companion species" Haraway thinks through human coexistence with dogs.[86] Within agrilogistic social space, cats stand for the ontological ambiguity of lifeforms and indeed of things at all. Cats are a *neighbor* species.[87] Too many concepts are implied in the notion of "companion." The penetrating gaze of a cat is used as the gaze of the extraterrestrial alien because cats are the *intraterrestrial* alien. Cats just happen. "Cats happen" would be a nicely ironic agrilogistic T-shirt slogan.

More to the point, consider bees again. Their symbiotic relationship with humans (let alone plants and the sexual facilitation thereof) could not be more obvious or more significant. Bees are moved en masse to where agrilogistics requires them; they are fed high-fructose corn syrup, a sick irony that could almost evoke a gallows-humor type of a laugh were it not so painful to think about. Monsanto's genetically modified, pesticide-coated seeds are causing Indian farmers to kill themselves and bees to die in their millions: the pesticides are fatal, but so is the modification of the plant structure itself, causing bees' intestinal walls to weaken. Global warming is forcing bumblebees north of their habitual pathways by about three miles a year, and they don't like it. The summer of 2014 was particularly bad, with about 42 percent of the U.S. bee population dying. The magic-bullet approach to getting rid of "pests" has resulted in this feedback loop: a range of pesticides called neonicotinoids are to blame. In response, it has not been very obvious to agrilogistics that improving the bees' conditions would help, because there is a general anthropocentric doubt that bees have conditions at all.[88] Instead, approaches such as Monsanto's war against the *Varroa destructor* mite infecting bees will only exacerbate the feedback loop. Axiom (3) (just existing is always better than any quality of existing) affects nonhumans too.

The agrilogistic engineer must strive to ignore the bees and the cats as best as he (underline *he*) can. If that doesn't work, he is obliged to kick them upstairs into deity status. Meanwhile he asserts instead that he could plant anything in this agrilogistic field and that underneath it remains the same field. A field is a substance underlying its accidents: cats happen, rodents happen, bees and flowers happen, even wheat happens; the slate can always be wiped clean. Agrilogistic space is a war against the accidental. Weeds and pests are nasty accidents to minimize or eliminate.

Consider the accident of an epidemic, which ancient Greek culture called *miasma*. Miasma is the second human-made hyperobject—the first was agrilogistic space as such, but miasma was the first hyperobject we noticed. You consider yourself settled and stable, although it would be better to describe your world as metastable: the components (humans, cows, cats, wheat) keep changing, but the city and the walls and the fields persist. You can observe miasmic phenomena haunting the edges of your temporal tunnel vision. You see them as accidental and you try to get rid of them. For instance, you move to America and start washing your hands to eliminate germs. Then you suffer from an epidemic of polio from which you had been protected until you started to police the temporal tunnel boundaries even tighter. This is the subject of Philip Roth's novel *Nemesis* and a good example of a strange loop.[89] The global reach of agrilogistics is such that antibiotic-resistant bacteria may now be found throughout the biosphere: "in environmental isolates, soil DNA . . . secluded caves . . . and permafrost," in "arctic snow" and the open ocean.[90] When you think it at an appropriate ecological and geological timescale, agrilogistics actually works against itself, defying the Law of Noncontradiction in spite of axiom (1).

The push to achieve constant presence in social and physical space requires persistent acts of violence, and such a push is itself violence.[91] Why? Because the push goes against the grain of (ecological) reality, with its porous boundaries and interlinked loops. Ecological reality resembles the shimmering, squiggly space of marks and signs underwriting the very scripts that underwrite agrilogistic space, with its neatly plowed lines of words, many of their first lines accounting for cattle—a lazy term as we have seen for anything a (male) human owns. Preagrilogistic "oral" social formats were not more present, as in the primitivist myth, itself a by-product of agrilogistics. Preagrilogistic social and conceptual space has far less to do with presence than agrilogistic space. Logocentrism—the idea that full presence is achievable within language, typified by the mythical utopian image of face-to-face communication—is an agrilogistic myth. This is why the deconstruction of logocentrism is a way to start finding the exit route.

Agrilogistic existing means just being there in a totally uncomplicated sense. No matter what the appearances might be, essence lives on. Ontologically as much as socially, agrilogistics is immiseration. Appearance is of no consequence. What matters is knowing where your next meal is coming from, no matter what the appearances are. Without paying too much attention to the cats, you have broken things down to pure simplicity and are ready for axiom (3):

(3) Existing is always better than any quality of existing.

Actually we need to give it its properly anthropocentric form:

(3) Human existing is always better than any quality of existing.

Axiom (3) generates an Easy Think Ethics to match the Easy Think Substance, a default utilitarianism hardwired into agrilogistic space. The Easy Think quality is evident in how the philosophy teacher in Stoppard's *Darkside* describes the minimal condition of happiness: *being alive instead of dead*.[92] Since existing is better than anything, more existing must be what we Mesopotamians should aim for. Compared with the injunction to flee from death and eventually even from the mention of death, everything else is just accidental. No matter whether I am hungrier or sicker or more oppressed, underlying these phenomena my brethren and I constantly regenerate, which is to say we refuse to allow for death. Success: humans now consume about

40 percent of Earth's productivity.[93] The globalization of agrilogistics and its consequent global warming have exposed the flaws in this default utilitarianism, with the consequence that solutions to global warming simply cannot run along the lines of this style of thought.[94]

Jared Diamond calls Fertile Crescent agriculture "the worst mistake in the history of the human race."[95] Because of its underlying logical structure, agrilogistics now plays out at the spatiotemporal scale of global warming, having supplied the conditions for the Agricultural Revolution, which swiftly provided the conditions for the Industrial Revolution. "Modernity once more with feeling" solutions to global warming—bioengineering, geoengineering, and other forms of what *Dark Ecology* calls *happy nihilism*—reduce things to bland substances that can be manipulated at will without regard to unintended consequences.

Planning for the next few years means you know where the next meal is coming from for some time. Who doesn't want that? And existing is good, right? So let's have more of it. So toxic and taboo is the idea of undoing axiom (3), one automatically assumes that whoever talks about it might be some kind of Nazi. Or that, given that we have seen population growth and food supply grow tougher, the one who doubts the efficacy and moral rightness of axiom (3) is simply talking "nonsense."[96] Nonsense or evil. Courting these sorts of reaction is just one of the first ridiculous, impossible things that ecognosis does. So much ridicule, so little time. Even more ridiculously, perhaps, we shall see that ecognosis must *traverse* Heideggerian-Nazi space, descend *below* it: through nihilism, not despite it.

It was based on increasing happiness, but within the first quarter of its current duration agrilogistics had resulted in a drastic *reduction* in happiness. People starved, which accounts for pronounced decreases in average human size in the Fertile Crescent. Agrilogistics exerted downward pressure on evolution. Within three thousand years, farmers' leg bones went from those of the ripped hunter-gatherer to the semisedentary forerunner of the couch potato. Within three thousand years, patriarchy emerged. Within three thousand years, what is now called the 1 percent emerged, or, in fact, the 0.1 percent, which in those days was called *king*. Desertification made swaths of the biosphere far less habitable. Agrilogistics was a disaster early on, yet it was repeated across Earth. There is a good Freudian term for the blind thrashing (and threshing) of this destructive machination: *death drive*.

Something was wrong with the code from the beginning. More happiness is better, such that more existing, despite how I appear (starving, oppressed), is better. We could compress this idea: *happiness as existing for the sake of existing*. A *for its own sake* that agrilogistics itself regards as superfluous or evil or evil because superfluous: nonsense and evil again, the way the aesthetic dimension haunts the Easy Think Substance. It sounds so right, an Easy Think Ethics based on existing for the sake of existing. Yet scaling up this argument unmasks a highly disturbing feature. Derek Parfit observes that under sufficient spatiotemporal pressure Easy Think Ethics fails. Parfit was considering what to do with pollution, radioactive materials, and the human species. Imagine trillions of humans spread throughout the Galaxy. Exotic addresses aside, all the humans are living at what Parfit calls *the bad level*, not far from Agamben's idea of *bare life*.[97] Trillions of nearly dead people, trillions of beings like the *Muselmänner* in the concentration camps, zombies totally resigned to their fate. This will always be absurdly better than billions of humans living in a state of bliss.[98] Because more people is better than happier people. Because bliss is an accident, and existing is a substance. Easy Think Ethics. Let's colonize space—that'll solve our problem! Let's double down! Now we know that it doesn't even take trillions of humans spread throughout the Galaxy to see the glaring flaw in agrilogistics. It only takes a few billion operating under agrilogistic algorithms at Earth magnitude.

There is a "very large finitude" in the shape of a specific, gigantic object (Earth) on which humans cooperate (and refuse to cooperate) with one another and with other lifeforms. There is also indeterminate futurity—how many future generations should we take into consideration? The combination of massive yet finite spatiality and massive and indeterminate time generates a very specific "game board" on which cooperation and its opposite play out. It seems clear in mathematics that a well-structured game board would ensure the best cooperation.[99] But the extremely minimal utilitarianism and ontology (Easy Think) implied by agrilogistics does next to nothing to determine the quality of the game board. The result is predictable: at any particular moment in the indeterminate time line it always seems better to destroy as much of the very large finitude as possible.

To avoid the consequences of the last global warming, humans devised a logistics that has resulted in global warming. Mary Daly is correct that we live in a death culture.[100] We Mesopotamians. In *A Scanner Darkly*, Philip K. Dick's novel about addiction and paranoia and the control society, the Latin name of the highly addictive and paranoia-inducing Substance D is *Mors ontologica*. Ontological death or, as someone in the novel says, "'Death of the spirit. The identity. The essential nature.'"[101] Robert Arctor gets completely fried by Substance D and enters a supposed rehabilitation center where he is recruited as walking death (bare life, aforementioned) to farm Substance D. The drug is, in fact, extracted from a tiny blue flower hidden amid gigantic fields of corn spreading to the horizon. The ironic inversion of the agrilogistic picture with its useful wheat and useless little flowers is stunning. Bare life harvesting ontological death, just executing an algorithm without a head: "'You can't make yourself think again. You can only keep working, such as sowing crops or tilling on our vegetable plantations—as we call them—or killing insects. We do a lot of that, driving insects out of existence with the right kind of sprays. We're very careful, though, with sprays. They can do more harm than good. They can poison not only the crops and the ground but the person using them. Eat his head.' He added, 'Like yours has been eaten.'"[102] Farming Substance D is evidently bad for the environment, and the state is well aware of the feedback loops, both inner and outer. . . . Taking it, farming it, suspecting one's brain while on it—all is relentless, mindlessly without laughter. Who is in charge of whom—the flower or the human? Nonhuman agency has been disastrously amplified by a human desire to "play" (Dick's term for drug consumerism), which has been in turn amplified and incorporated into the control society.[103] The deadly serious play of oppression exemplified in the world of Substance D is absolutely the opposite of coexistence otherwise than agrilogistics: as the Third Thread will show, this looks more like *playful seriousness.*

Curiously, while it rots your brain, Substance D makes you surprisingly compassionate toward nonhumans. In the midst of the absolute nightmare of state-controlled death-in-life, some kind of care evolves, though it looks like decadence, like Nietzsche weeping with the whipped horse. Perhaps this is how true progress looks to a society hell-bent on speed: like the absurd number of hours it takes for a group of "heads" to remove a shard of glass from the stomach of a cat without hurting her.[104] Fumbling for the exit route is still occurring, a curious phenomenon we shall explore in the next thread.

Nature = agrilogistics. At the end of *A Scanner Darkly*, Robert Arctor is shown mountains that lie beyond the farms of Substance D: "'Mountains, Bruce, mountains.'" [105] It is an absolutely circular, flat, tautological description in which the simple phrase the manager uses is echoed exactly by the now mindless "Bruce" (Arctor renamed by the rehabilitation center). The echoed phrase echoes itself, cycling in a loop fed back to Bruce, who is a mere cipher, barely life, not even owning his own name, just repeating the phrase to the manager like a mirror. As if the manager were introducing the mountains to Bruce, and Bruce to the mountains: a deadly sincere chiasmus. Mountains, Bruce, phrases—all are substances without qualities, like the mysterious Substance D itself, whose immediate psychophysical effects appear absolutely nonexistent. Substance D is the drug of meta:

going out of one's mind on it consists of wondering at higher and higher levels whether one is going out of one's mind, dissociating to the point where one could seem to be investigating oneself as an agent would investigate a suspect. Purged of its playful blueness and little-flowerness according to the logic of the "active ingredient," Substance D is the Easy Think Substance transmuted into an addictive drug: serious play. People assume it is entirely synthetic, but it is in fact "organic," the product of human interactions with nonhumans via agrilogistics.[106] *Organic*, a rich and serious term for a rich and serious circularity without play or excess or brokenness or deviance: mountains, Bruce, mountains. A zombie substance for zombie human substances.

Don't we have here, crushed together in the frightening mixing bowl of Dick's spare prose, the Cartesian manifold stripped of comforting references to religion? On the one hand, absolute paranoia—as I wonder whether or not I exist, I can't help wondering whether I might be the puppet of some all-powerful but invisible demon.[107] On the other hand, absolutely bland extension, pure substance without end. A man without a head looking at himself looking at himself: mountains, Bruce, mountains. As if the point of that phrase were simply to make more of itself, like the farms of Substance D or Marx's scary encapsulation of capitalism in a tellingly similar phrase, *M–C–M'*, where money loops through capital and multiplies. Pure survival without quality, based on fear, generating people who can't tell whether or not they are people working on objects they can't tell are objects. *Mors ontologica* indeed. Which is why ontology is a vital part of the struggle against agrilogistics.

Mountains, Bruce, mountains: in other words, Nature, a substance "over there," underneath, just round the corner, despite appearances, out back, behind the surface, comfortingly present, endless, normal, straight. Agrilogistics spawns the concept of Nature definitively outside the human. The normative concept of Nature, telling you what's "in" and what is "out," as surely as a jaded fashion magazine, is deeply troubled. Normative Nature simply can't cover absolutely everything because Nature depends on specifying the unnatural. But this is just what we moderns are incapable of doing in advance of the data. The concept Nature is a flicker of resistance to the oncoming metal army of industrialization, like a fake medieval sword made of rubber. A fake medieval sword that heightens the fire risk in California's Yosemite National Park: John Muir, architect of the parks and believer in Nature, favored the growth of trees that covered the slopes in attractive (and flammable) swaths of dense green, to the chagrin of the Native Americans.[108]

The rhetoric of what I have elsewhere called ecomimesis is typified by a Nature speak that tries to straighten out a loop.[109] The core of ecomimesis is a sentence such as "As I write this, I am immersed in Nature." Ecomimesis tries to fuse the layer of narrative and the layer of narration, creating a paradoxical loop about whose paradoxical and loopy qualities ecomimesis is perpetually in denial. The denial within ecomimesis is a symptom of the larger loop of whose machination ecomimesis is a small, human-scaled, "lived experience" region.[110] Its job is to flatten out the inherent twist in a chiasmus, to make the twist into a pure circle, "an insect that clacks and vibrates about in a closed circle forever."[111]

Closing the circle is impossible. Even a circle is a *circling*, a circulation that implies an inherent movement, a constant deviation from the integral (pi, impossible to compute completely, yet thinkable). A circle is thinkable yet impossible to execute, the very opposite of agrilogistics, which by contrast is pure execution without a head. Even a circle is twisted. Attempts to straighten things are violent; they never work perfectly because they are "doomed." When we hear the phrase *Mountains, Bruce, mountains* and its pure echo, we are haunted by something, an excess in the very doubling, the very circularity, the *invagination* that turns things inside out. Something lopsided and broken, crying with pain, a shard of glass in its stomach, stuck between the inside and the outside of a house, a human dwelling (Greek *oikos,* hence *oik*-onomy, *oik*-ology).[112] A cat stuck between inside and outside: an *intraterrestrial alien* haunting the supposed pure circularity of Nature and human (mountains, Bruce). The edge of a circle is a deviation. The edge of a system such as agrilogistics is a fold, a twist.[113] The edge is not absolute.

In this sense, the concept Nature isn't only untrue; it's responsible for global warming. Nature is defined within agrilogistics as a harmonious periodic cycling. Conveniently for agrilogistics, Nature arose at the start of the geological period we call the Holocene, a period marked by stable Earth system fluctuations.[114] One might argue that Nature is an illusion created by an accidental collaboration between the Holocene and agrilogistics: unconscious, and therefore liable to be repeated and prolonged like a zombie stumbling forward. Like Oedipus meeting his father at the crossroads, the cross between the Holocene and agrilogistics has been fatally unconscious.

Nature is best imagined as feudal societies imagined it, a pleasingly harmonious periodic cycling embodied in the cycle of the seasons, enabling regular anxiety-free prediction of the future. Carbon dioxide fluctuated in a harmonious-seeming cycle for twelve thousand years—until it didn't.[115] We Mesopotamians took this coincidence to be a fact about our world and called it Nature. The smooth predictability allowed us to sustain the illusion. When we think of nonhumans we often reminisce nostalgically for a less deviant-seeming moment within agrilogistics, such as fantasies of a feudal worldview: cyclic seasons, regular rhythms, tradition. This is just how agrilogistics feels—at first. The ecological value of the term *Nature* is dangerously overrated, because Nature isn't just a term—it's something that happened to human-built space, demarcating human systems from Earth systems. Nature as such is a twelve-thousand-year-old human product, geological as well as discursive. Its wavy elegance was eventually revealed as inherently contingent and violent, as when in a seizure one's brain waves become smooth.[116] Wash-rinse-repeat the agrilogistics and suddenly we reach a tipping point.

The Anthropocene doesn't destroy Nature. *The Anthropocene is Nature* in its toxic nightmare form. Nature is the latent form of the Anthropocene waiting to emerge as catastrophe.

Agrilogistics is a strange loop because its very attempt to smooth out the physical world and to smooth out anxiety doubles down on that physical world and on anxiety itself, just like washing your hands forces bacteria to adapt. Why did this strange loop emerge? How can we think this emergence? It would be going against the implicit temporality of loops to assert, as so many do, that there was an origin point, exactly there, exactly then, constantly present in a definable archive.[117] Such an assertion is recursively part of the very agrilogistic schema we are attempting to explain. Instead of looking for an origin point then, we must think ecologically. We must examine how an existing state of affairs (ecosystemic degradation resulting from global warming) interfaced with an existing state of affairs (human psyches). Moreover, we must think each state of affairs as entwined with one another and as consisting of nested loops of other states entwined with one another: humans within ecosystems, thoughts within brains. A nest of vipers.

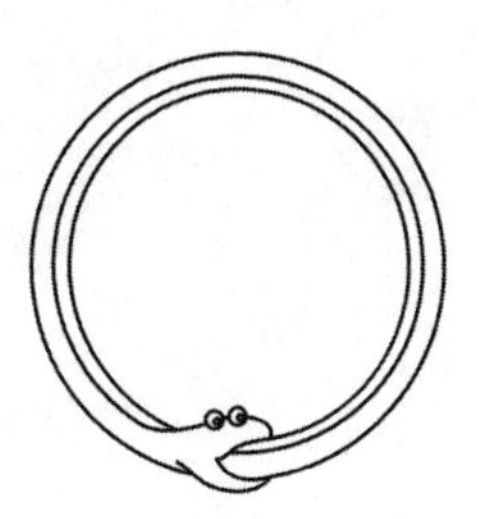

THE SECOND THREAD

The Emmets Inch & Eagle's Mile
Make Lame Philosophy to smile

What is happening?

Lame Oedipus tried to elude the prophecy that he would kill his father and marry his mother. In so doing he reinforced it. *His attempt to escape the web of fate was the web of fate.* Oedipus' ancestor Cadmus founded Thebes by slaying a gigantic serpentine dragon. Then Cadmus sowed the dragon's teeth and out they sprang, the Spartoi, the sown ones, fully armed. They selectively bred themselves like wheat by fighting until only a handful was left, ready to defend the agricultural city-state. A field grows war on the back of a slain nonhuman because agriculture is already weaponized. As the book of Joel observes, swords into ploughshares means you can convert ploughshares into swords (Joel 3:10). A brief visit to an agricultural museum will demonstrate how tanks were ploughs first.[1] Admittedly the dragon wasn't as bad as the Lernaean Hydra that Hercules had to confront. In that case, two heads grew back where one used to be, a positive feedback loop. Somehow Hercules slew this female monster, which technically couldn't be slain. Perhaps for his benefit the rules had been suspended, as if to mark, silently, the structural impossibility of agrilogistics.

In his anthropocentric reading of Oedipus, Freud forgets the Sphinx, the ambiguous female monster lurking on the edge, vague nothingness: the monster that haunts the boundary of agrilogistics.[2] The Oedipus complex is agrilogistic insofar as it deletes this nonhuman. There is nothing but Oedipus and his parents, Oedipus who thinks he acts autonomously, exemplifying the agrilogistic meme *We came from ourselves*. Oedipus kills feminized nothingness, his logic causing the Sphinx's ambiguous image of "man" (four legs at dawn, two legs at noon, and three legs at eve) to collapse into noncontradictory consistency.

Oedipus the lame philosopher reduces the riddle's ambiguous appearance to an Easy Think Substance that doesn't depend on accidental properties. Riddles assert links between the human and the nonhuman, links that are intrinsically mysterious. The "Look! I see something" that begins a Koyukon Athabascan riddle marks out nonhumans without parceling them into bite-size pieces.[3] Likewise, Old English riddles talk about onions, bookworms, oysters. Riddles are funny because they exploit an irreducible gap between what a thing is and how it appears.[4] Riddles are realist because things are riddles. Oedipus' answer "Man" is not so funny, closing down the riddling suspension of the many-legged being, legs that reveal a metamorphic time, not a linear succession of predictable now points. *Man* is a spoiler, not a punch line.

Rather than murder the Sphinx in a matricidal act reminiscent of some earlier phase of agrilogistic misogyny, Oedipus' cool wits cause the Sphinx to kill herself, perhaps an even more misogynistic act. As if the violence of agrilogistics is erased by a more "progressive" form of agrilogistic violence, a free-floating reason detached from its physical basis in the body, enabled by the lowering of anxiety and the fullness of the belly.[5] While agrilogistics 1.0 may have found Oedipus' defeat of the Sphinx sacrilegious, quantizing the riddle to an Easy Think Substance reveals something at the core of agrilogistics as such. Oedipus' *hamartia* is his reason, and his hubris is to use his wits to command everything, as if reason could shrink-wrap the universe. The prime example of the loop we explored in the First Thread, Oedipus is the criminal he is seeking, responsible for the miasma gripping Thebes.

There is a drive within agrilogistics for constant upgrades. In the very myth of the Sphinx we discover an older layer of agrilogistic code supplanted by new "civilizations": what Oedipus battles has the look of an Assyrian angel, a female lamassu mashed up with a Fury, bringer of miasma. The Sphinx is half-buried evidence of the long agrilogistic retreat now playing out in the drought-ridden Central Valley of California. The half-burial of older agrilogistic layers in the Oedipus myth is itself fascinating, as if the myth as such were a many-headed hydra, a loop that persists despite continued efforts to stamp out all the weeds, a founding narrative that has to be retold over and over again—*because it never fully happened*. The uncanny mashing of the lamassu and the Sphinx and the Fury points to something growing between the cracks of agrilogistics itself. It points to what *Dark Ecology* will call the *arche-lithic*, a primordial relatedness of humans and nonhumans that has never evaporated. Bruno Latour argues that we have never been modern. But perhaps *we have never been Neolithic*. And in turn this means that the Paleolithic—adore it or demonize it—is also a concept that represses the shimmering of the arche-lithic within the very agrilogistic structures that strive to block it completely. We Mesopotamians never left hunter-gathering mind.

Fully to understand the arche-lithic, however, we must traverse the founding myth and its logics. Isn't it interesting how Sophocles himself gets wind of the need to assert and then reassert the agrilogistic in the Theban plays themselves, as if agrilogistics were betraying through him a compulsion to repeat? Consider how in the aftermath of the misogynistic battles against monsters there arises the phenomenon that inspires the terrifying Second Chorus in *Antigone*, the story of Thebes in the postoedipal generation. Anne Carson's adaptation is haunting:

> Many terribly quiet customers exist but none more
> terribly quiet than man
> his footsteps pass so perilously soft across the sea . . .
> and every Tuesday

down he grinds the unastonishable earth
with horse and shatter . . .
Every outlet works but one
: Death stays dark.[6]

The key motif? *Plowing*. Sophocles seems to be indicating a level resembling Earth magnitude, at which the quotidian activity of plowing can be seen as physically, psychologically, and philosophically *disturbance*. The Chorus states, "Polla ta deina k'ouden anthrōpou deinoteron pelei." This is translated by Carson as "Many terribly quiet customers exist but none more / terribly quiet than man." The quietness here is evidence of something sinister hiding in plain sight. Heidegger liked to translate *deina* and *deinoteron* as "uncanny": of all the many uncanny beings out there, none is more uncanny than man.[7] Perhaps *uncanny* doesn't get at how *deina* suggests *disturbing*, including the disturbance of the uncanny itself. What is most uncanny about human being is its attempt to rid the world of the uncanny. Or, and this is putting it in its most ecological register: human being disturbs Earth and its lifeforms in its desperate and disturbing attempt to rid itself of disturbance.

Weird essentialism. Can we think the Sphinx without packaging her as Easy Think Man? Can we think the Sphinx within the contemporary correlationist consensus? This is what we need to try.

The correlationist often assumes that because a thing is real insofar as I have to open its refrigerator door to see if it exists, that means *it doesn't really exist*. The idea that I am some George-Bush-like Decider who calls the shots on what exists is directly related to the Sixth Mass Extinction Event. *I'm the Decider* means *I get to have what I want, when I want it*. A thing is a blank screen. Although it claims to have transcended ontology, correlationism reproduces the Easy Think Substance. In a way it makes the Easy Think Substance more toxic: a thing is anything I can do to it . . .

Earth isn't just a blank sheet for the projection of human desire: the desire loop is predicated on entities (Earth, coral, clouds) that also exist in loop form in relation to one another and in relation to humans. We are going to have to rethink what a thing is. We require a Difficult Think Thing. Our exploration of weirdness and its suppression by agrilogistics suggests that the weird might be a helpful ontological category. We are going to have to think things as *weird*. That I claim humans exist and made the Anthropocene by drilling into rock does indeed make me an essentialist. However, if we must attune to the Difficult Think Thing, such a thing wouldn't cleave to the Law of Noncontradiction, agrilogistic axiom (1). Which, in turn, implies that while beings are what they are (essentialism) *they are not constantly present*. Demonstrating this would constitute a *weird essentialism* in the lineage of Luce Irigaray, whose project has been to break the Law of Noncontradiction so as to liberate beings from patriarchy.[8]

It's quite evident that from the beginning ecognosis was installed within the weird essentialism exemplified by French feminism. In the excellently named *Let's Spit on Hegel*, Carla Lonzi writes: "The women's movement is not international but planetary."[9] Or consider the thought of Françoise d' Eaubonne, coiner of the very term *ecofeminism*.[10] Yet the ritual exclusion of essentialism from serious consideration in theory class has made attending to ecofeminism almost unthinkable within that setting. The biocentric ecofeminists in 1970s America began to retreat from their witchy ways into Habermasian consensus speak, cowed by the antiessentialists. And early 1990s ecocriticisms were forms of antitheoretical Nature speak, reactions to modernity that reinforce modernity.[11]

As a performance of not seeming an idiot in theory class, one is obliged to convey something like, "Well of course, I'm not an *essentialist*" (make disgusted face here). Compare the ridicule that greets the idea of creating social spaces that are not agrilogistic (so not traditionally capitalist, communist, or feudal). Such reactions are themselves agrilogistic. Both assume that to have a politics is to have a one-size-fits-all Easy Think concept. If you don't, you are called a primitivist or an anarchist, both derogatory terms, and deemed unserious. Or you want to regress to some utopian state that "we couldn't possibly even imagine." "Of course, I'm not advocating that we *actually* try a social space that includes nonhumans in a noncoercive and nonutilitarian mode." Or its inverse, ridiculing "civilization": insisting that humans should "return" to a preagrilogistic existence (John Zerzan, archivist of the Unabomber Ted Kaczinski). "Eliminate the evil loops of the human stain. Anyone with prosthetic devices such as glasses is suspect."[12] Once one has deconstructed civilization into agrilogistic retreat it is tempting to think this way. But imagine the Year Zero violence of actually trying to get rid of intellectuality, reflection, desire, whatever we think is a source of evil, so we can feel right and properly ecological. The assertion that this is a problem to do with "domestication"—which is how Zerzan and others frame it—avoids the genuine agrilogistic problem. "Domestication" is a term from some kind of fall narrative: once upon a time, we let things be wild, but then we took some into our homes and unleashed evil. Neanderthals lived in homes. Primates make beds of leaves. Dogs were fused with humans hundreds of thousands of years ago. "Domestication" is a canard that is itself agrilogistic, straight out of a theistic fall narrative.

How Mesopotamian. It is as if, whenever an origin or an end to the agrilogistic program is sought, we run into a self-imposed limit that is itself agrilogistic. The very idea of points of origin is an agrilogistic hallucination. For what is agrilogistics itself but the blind execution of a program in which *my* blind execution of key turnings in ignitions betrays my ignorance of the hyperobject of which I am a component, the human species? What needs to be asked, rather than *How did it all start?* is *Why on earth is this execution blind?* Why has it been so since the program started to run 12,500 years ago? You can't decide to execute instructions blindly.

Isn't it more plausible to hold, like Julian Jaynes, that *we were told to do it*? In the fascinating and strange *The Origin of Consciousness in the Breakdown of the Bicameral Mind* Jaynes argues that humans once considered their thoughts to emanate from beings outside themselves: hence "bicameral," which literally means "consisting of two chambers." According to this argument, voices taken as commands told us to execute the agrilogistic program.[13] In a sense, the voices are viruses, alien entities in our heads whose slavish execution we still see around us in churches and temples, houses of gods that are not simply hangovers from early agrilogistics but direct and persistent expressions of it, as viruses persist on tabletops and indeed in our brains until something switches them on. After all, God does seem to behave this way in Genesis: go forth and multiply; dust shalt thou eat. Jaynes's argument is more plausible than Yuval Noah Harari's assertion that it was all about "imagination" creating "fictions" to dupe people, and unlike him, Jaynes requires no unproven beliefs.[14]

If the very question of inside and outside is what ecology undermines or makes thick and weird, surely this is a matter of seeing how ecosystems are made not only of trees, rock formations, and pigs (seemingly "external" to the human) but also of thoughts, wishes, fantasies (seemingly "inside" our human heads)? And isn't this at least plausible since it's quite logical to argue, as I shall do soon, that thoughts themselves are independent entities, reducible neither to brain nor to mind—just as pigs are independent entities reducible neither to parts of pigs or prepig ancestors or the ecosystems of which they are members?

So isn't it better to argue against the idea that language emerged one fine day and thus that auditory hallucinations were attributed to voices of gods? But not from the point of view of rejecting the very idea of hallucinations and gods, of nonhuman entities not obviously made of scientistic objects. Isn't it better and more energy efficient, cognitively speaking, to claim, as I shall be, that *we have always been hallucinating* and that what happened was not entirely internal to the human (mind or brain) or external to the human (environment), but was rather a weird entwined fusion of both, a twisted turn of events symbolized by Urðr, the Norn of causality?

Perhaps then what you experience in a desertifying ecosystem where your next meal begins to evaporate is a hallucination that reassures you, like how a voice talks to the driver in the midst of a bad car accident, that you can cope, that if you just grip the wheel so, or set up camp over there, it will be all right. Aside from our demonization of them as schizophrenia, a demonization that itself is an agrilogistic symptom, we all hear these voices, and frequently.[15] Perhaps agrilogistics is a thought-virus taken to have been a definitive command with the retroactive theistic blindness of 20–20 agrilogistic hindsight. In short, what Derrida calls *logocentrism* is evidence of the virus having been taken as just such a command. Taking a phrase, literally a virus ("Save the Whale!" "Reds Under the Bed!" "Enjoy Coke!" "Our Father, who art in Heaven . . . "), to be a unit meaningful unto itself rather than as a wriggling linguistic worm, a *turn* (*urth*) of phrase otherwise known as a *trope*.

We hear a compelling phrase and execute its supposed command. It happens all the time. It happened to Neanderthals. It happens in discos. Put your hands in the air and wave them like you just don't care. Language as phatic proclamation of existence, of being "here": Say a hip, hop, hip to the hippety hip hip hop and you don't stop the rock, say up jump the boogie to the rhythm of the boogie da beat.[16] That's OK. What is not OK is ascribing a telos to the phrase. What is not OK is the very origin-ization of the phrase, taking it as the god's command from time immemorial and for time immemorial, saying to ourselves that this phrase and this phrase only is the one true phrase for all time. Blind execution means: I am stuck in a Turing machine whose stopping point I can't discern. Who or what can, every time? I can't help it. An endless groove. Not human, not natural, just a virus, a planetary earworm of which we have all become vectors. And why? Because the endless groove is so soothing. The idea that it originated somewhere does not eliminate its endlessness, but is rather the retroactive justification for its pleasure: *Once upon a time* . . . or *Since the dawn of time, Mankind* . . .

We feel as if we have never stopped dancing in the agrilogistic disco, the longest-running pop tune in human history. This feeling is a side effect of the tune itself. The very question, *How did we enter the disco, and when?* is the kind of thing you end up thinking in a disco like that. The question of origin is bound up with the compulsion to execute blindly. Blind execution suppresses, yet expresses in its very form its frantic nonstopping, a bedrock anxiety.[17] Agrilogistics speaks with the reassuring anxiety of Dory in *Finding Nemo*: *Just keep swimming, just keep swimming, just keep swimming* . . . [18] Now that its blind execution covers a sufficient extent of the planet's surface, we confront the initial anxiety, never dispelled: the environmental catastrophe of global warming from which we were swimming.

Yet we were not always dancing in the agrilogistic disco. There *is* an "origin." How to think it? Which is to say, *How to have a realism not subject to the metaphysics of presence?* We have to tell the story differently. The notions of "origin" and "point" involve questions about how to think time, knowing what we know of geology.

Concentric temporalities. One of the things we need to rethink weirdly is time. If future coexistence includes nonhumans—and *Dark Ecology* is showing why this must be the case—it might be best to see history as a nested series of catastrophes that are still playing out rather than as a sequence of events based on a conception of time as a succession of atomic instants. We can think these nested sets as ouroboric, self-swallowing snakes; it isn't surprising that many first peoples imagine the outer rim of reality as an entity like Jörmungandr, the Norse serpent who surrounds the tree of the universe.

Why is it better for nonhumans this way? At the temporal scale of global warming, the human as historicity—the correlator that makes things real by bringing history to the picnic of data—becomes inoperative. Geological eras are *nested* catastrophes. Consider the air you are breathing in order to stick around for the next sentence. Oxygen is an ecological catastrophe for the bacteria that excreted it (starting about 2.3 billion years ago). The Anthropocene is a loop within a much larger loop we could call the Bacteriocene. The Bacteriocene and its oxygen are happening *now*, otherwise I would be writhing on the floor rather than typing this sentence. The Oxygen Catastrophe was not an event in atomic time. Surrounding the Bacteriocene there is the Cyanidocene, the moment of the strange dance of death-and-life between nucleic acids, proteins, and hydrogen cyanide polymers; cyanide itself having very likely formed as the result of a further cataclysm, a gigantic comet or asteroid impacting Earth.[19] *The Cyanidocene is happening now*—otherwise I would be a puddle of chemicals.

The loop of the Cyanidocene exists within an even more encompassing one in which organic molecules began to replicate, a loop we could call the Mimeocene (from the Greek *mimēsis*, copying), acknowledging the emergence of self-replicating molecules. Going even wider, we discover what we could name the Haemocene (from *haimos*, Greek for *iron*). Earth's liquid iron core began to spin around its solid center, emitting an electromagnetic shield that enabled life to evolve by protecting organic molecules from solar rays. This too is happening now, otherwise I would be a charred corpse. The loops are not hermetically sealed from one another, which is why they happen at all. They are happening now. So the Anthropocene is a small region of the Bacteriocene, which is a small region of the Cyanidocene, and so on. These temporality loops all happen in a nowness I cannot reduce to an atomic point of whatever size.

What is the dynamic of loop formation? On the view of weird essentialism, things are inconsistent rather than constantly present: to be a thing is to have a gap between what you are and how you appear. Thus any attempt to resolve the intrinsic inconsistency of a thing creates loops that scale up to catastrophes. Weird essentialism defines an *event* as twisted novelty emerging out of a weird distortion of its conditions. In trying to get rid of toxic oxygen, bacteria inadvertently brought about the conditions in which I am now breathing. Inconsistency is why along with bacteria there arose viruses, because the boundary between a living and a nonliving thing isn't thin and clear. Attempts to create consistency—for a single-celled organism to maintain itself in a metastable state, for instance—result in parasites that exploit structural weaknesses in their hosts. Why are there viruses at all? A virus exploits inherent inconsistencies in cells not unlike the inherent inconsistency of logical systems. A logical system is true on its own terms if it can be forced to talk about itself—go into a loop—and say contradictory things such as "This sentence is unprovable" (a tiny version of the Gödel sentence). I take this to indicate that entities (the *Principia Mathematica*, Mexican heather) are *inherently inconsistent*. A catastrophe is a twist—the Greek means "downward turn"—in the already twisted spatiotemporal fabric of an existing catastrophe.

Fuzzy temporalities. Temporality structures such as the Anthropocene are fuzzy and not atomic because things in general are fuzzy and not atomic. A human being is an ecosystem of nonhumans, a fuzzy set like a meadow, or the biosphere, a climate, a frog, a eukaryotic cell, a DNA strand. We might begin to think these things as wholes that are weirdly *less* than the sum of their parts, contra the usual rather theistic holism where the whole is always greater than its parts. There is literally more non–Tim Morton DNA in Tim Morton than there is Tim Morton DNA, as a condition of possibility for Tim Morton's existence. In order to allow these fuzzy sets to exist, logic must relax its grip on Law of Noncontradiction, the rule whereby Tim Morton must coincide with Tim Morton in order to exist.

We had certainly better do so if we are going to think symbiosis. If we don't, we end up with a state of affairs in which there are either no lifeforms, since they are all composed of symbiotic communities to a greater or lesser extent (from lichen to the stomach's microbial biome), or in which symbiosis is fundamentally impossible, since the boundaries between individuals must be thin and rigid.[20] The fact of symbiosis requires that we think a *weird essentialism*: there are distinct lifeforms insofar as a frog isn't a peach; but they are not your grandfather's "distinct"—well, not my grandfather's anyway.

Likewise, if the boundary of the Anthropocene were thin or rigid we would encounter Zeno's paradoxes as we approached it. We could subdivide the approach into infinitesimal temporal parts and so never reach the boundary. It would strictly be impossible to cross the boundary, in the same way that you can't be "in the doorway" if you believe that you are either inside or outside a room—that is, if you believe in the Law of the Excluded Middle, which is a consequence of the Law of Noncontradiction.

Why the obsession with impossibly tidy boundaries? Nietzsche argues that logic itself is "the residue of a metaphor."[21] Despite the concept of logic "as bony, foursquare, and transposable as a die," logic is saturated with fossilized social directives. Hegel had an inkling of this when he distinguished between logic and thinking, that is to say between the mind's movement and the manipulation of preformatted thoughts. Nietzsche asserts that language is caught up in the caste system—and let's not forget that the caste system is a direct product of agrilogistics. With uncanny insight, Nietzsche himself seems to confirm this when he then asserts that logic as such is a symptom of caste hierarchies. Without doubt, these hierarchies oppress most humans. But don't class distinctions depend, as Cary Wolfe has argued, on a deeper *speciesism* that separates the human from the nonhuman, the better to oppress the nonhuman?[22] The human caste system, itself a product of agrilogistics, sits on top of a fundamental caste distinction between humans and nonhumans, a founding distinction wired into the implicit logic of agrilogistics.

Recall, furthermore, that some of the most common words for thinking and apprehension—*gather, glean*—derive from agriculture.[23] What is required is no less than a logic that is otherwise than agrilogistic. A logic that is fully eco-logical. If you want ecological things to exist—ecological things like humans, meadows, frogs, and the biosphere—you have to allow them to violate the logical "Law" of Noncontradiction. Let us explore this.

Zeno's paradoxes are not the only ones that arise if our boundaries are too tidy. Imagine a meadow—it's filled with grasses and flowers, bees are buzzing around, there are some trees, some water is flowing, small mammals are creeping about, butterflies land on petals. I remove a blade of grass from the meadow. Is there still a meadow? Why yes. I remove another blade of grass. There is still a meadow. Then another. And another. At every stage I can answer that there is a meadow. By now I have removed all the grass. I have a huge patch of dirt, and the butterflies have gone somewhere else. According to my logic, there is still a meadow! So, because I adhere to the Law of Noncontradiction, there is no such thing as an *actual* meadow—because it might as well just be a huge patch of dirt. Perhaps I can turn it into a parking lot now.

Let's try it in reverse. I plant a single blade of grass on the bare patch of dirt. That doesn't make a meadow. Let's plant another one. Still no meadow. I go on and on. Soon I have planted tens of thousands of blades of grass. According to my logic, which is correct at every step, there is still no meadow! Now I can see butterflies flitting about and voles clinging to the longer stalks of grass. Yet, according to my logic, there can't be real meadows. Why? Because if there were a real meadow I would have contradicted myself when I concluded, correctly on my own terms, that there was not a meadow. Or vice versa: every time I stopped to check whether my grass removal removed the meadow, I would be contradicting myself to say that there was no meadow when, according to my logic, the meadow was intact.

There is no single, independent, definable point at which the meadow stops being a meadow. So there are no meadows. They might as well be parking lots waiting to happen. And since by the same logic there are no parking lots either, it doesn't really matter if I build one on this meadow. Can you begin to see how the logical Law of Noncontradiction enables me to eliminate ecological beings both in thought and in actual physical reality? The Law of Noncontradiction was formulated by Aristotle, in section Gamma of his *Metaphysics*. It's strange that we still carry this old law around in our heads, never thinking to prove it formally. According to the Law of Noncontradiction, being true means not contradicting yourself. You can't say *p* and *not-p* at the very same time. You can't say *A meadow is a meadow and is not a meadow*. Yet this is what is required, unless you want meadows not to exist.

We have seen how contemporary thought shows how beings no longer coincide with their phenomena. Things become misty, shifty, nebulous, uncanny. The spectral strangeness that haunts being applies not only to single lifeforms—a vole is a not-vole—but also to meadows, ecosystems, biomes, and the biosphere. The haunting, withdrawn yet vivid spectrality of things means that there can be sets of things that are not strictly members of that set, such as a meadow, and this violates Bertrand Russell's prohibition of paradoxical sets that contain members that are not members of them. Meadow-type sets resemble Georg Cantor's transfinite sets. Transfinite sets are impossibly larger than infinity as previously thought, and strictly impossible to see or count. Yet in Cantor's brilliant diagonal proof one can see with one's naked eyes the crack in the real, the gap in mathematizing things. That is what is uncanny—it is as if you can see the gap, the nothingness. Imagine a grid on which are arranged all the rational numbers, in sequence from the first to the whateverest. Appearing down the diagonal of this incredible list is a weird, monstrous, deviant number that literally slants away from the others. It is not in the infinite set of rational numbers, by definition! No wonder there is a legend that Pythagoras, worshipper of the sacred integer, drowned Hippasus, who had discovered irrational numbers.

Transfinite sets contain contradictory sets of numbers. There is an irreducible gap between the set of real numbers and the set of rational numbers—Cantor drove himself crazy trying to find a smooth continuum between the two. This drive to find a continuum is a hangover from the Law of Noncontradiction. Contemporary thinking cannot cleave to a logic that assumes that things are rigid and brittle, that things do not contradict themselves. Irigaray observes that women's speech is alogical: "she steps ever so slightly aside from herself" with "contradictory words."[24] Irigaray asserts that women's bodies are "neither one nor two."[25] Irigaray does something we are now doing: moving from language to being somewhat fluidly. Women's contradictory speech happens because women are contradictory. We can extend what Irigaray argues about women to include men, spoons, quasars, and meadows.

When we think about taking meadows apart or building them up blade by blade, we confront *the Sorites paradox*, the problem of heaps: what constitutes one if you cleave tightly to noncontradiction? Since a human is a heap of things that aren't humans, just as a meadow is a set of things that aren't meadows, such as grasses and birds, either ecological and biological beings don't really exist or there's a malfunction in the logic we have rather uncritically inherited from Aristotle. A malfunction, moreover, that is beginning to distort political decisions at scales appropriate for thinking global warming. If we relax our grip, we can allow for sets of things that don't sum to a whole, and this just is what we have when we think geological temporality as a series of nested sets of catastrophes.

Catastrophes resemble meadows. First, they are physical and have a special experiential property. A catastrophe is what you experience when you are caught in a loop. Wouldn't knowing ourselves as a species be like waking up to find that we weren't floating in a void, but were inside the stomach of a gigantic worm, like Han Solo and Princess Leia in *The Empire Strikes Back*? Outside the loop, the perturbation has no significance. From the point of view of the entropy at the end of the universe (assume it has the Muppet-like ability to talk), who cares about the Anthropocene? From the point of view of

Earth's electromagnetic shield, who cares what happens to some slimy proteins in the newly formed oceans? Who cares, say the proteins, if some of us start to unzip ourselves, a process that if DNA and RNA were people might be construed as a death wish that results in the reproduction of the very entity that was unzipping?

Secondly, catastrophes defy noncontradiction. Catastrophes are receptacles like Plato's *chora*: they are weird places that don't have thin rigid boundaries.[26] Another one pops open inside an existing one when some property of an existing system begins to go into a strange loop, giving rise to another receptacle. The twist in the loop enables it to differ from its surroundings, namely the consequences of a more ancient and widespread loop.

Thinking temporality structures in this manner would clarify the fraught debates within stratigraphy, the study of geological periods via geological strata. There is a debate as to when the Anthropocene started: some say 1800, some 1945, and some say earlier, perhaps when Europeans began colonizing non-Europe. The debate often hinges on how much the debaters cleave to noncontradiction. If they cleave tightly, there must only be one date. And if they add to this a metaphysics of presence, 1945 begins to look right because the data spike is so vivid and compelling, 1945 being the date at which begins the Great Acceleration of the processes unleashed by fossil fuel burning. But saying that 1945 was the start of the Anthropocene is like saying that a person was shot when it became obvious that she was a corpse. Instead, we might be able to answer yes in different ways to 1945, 1800 or some earlier date. But the Anthropocene did not start 1.3 million years ago. The Anthropocene is temporally fuzzy, not absolutely indeterminate.

The Anthropocene is an event within agrilogistic space, which is quite evidently still happening now. Since agrilogistics requires human vectors, something in the structure of the inner logic of agrilogistics must mesh with the human desire to eliminate anxiety. We are now in a position to examine how human minds get behind a scheme that to an extraterrestrial would indeed appear to be a catastrophe, a downward turn of events. The nice thing about concentric temporalities defined as catastrophes is that since they are ongoing, we might be able to do something about them. Now that we know temporalities are fuzzy and ongoing, perhaps we are ready to explore a temporality that I shall be calling *the arche-lithic*. For it is my contention that this temporality is very much with us right now and that it provides a way to think how to unwind the catastrophe of agrilogistics.

Paleolithic, Neolithic, arche-lithic. Three names, two in common circulation, naming different relationships between humans and stones. Do humans call the shots in that relationship? For Lévi-Strauss the difference between horses and axes is that while horses reproduce on their own, axes can't: axes are completely subsumed within human meaning.[27] But are they? Who is manipulating whom? A human carves a stone to make an axe. She is following directives issued by the stone, the cutting tool, the tree, the wooden handle—and the hide and flesh and bone of the animal the axe meets. Whoever uses the axe responds to similar directives.[28] Humans are sensitively susceptible to stones and flesh and wood, whether or not they are seen as alive.

Who is host and who is parasite? Not to mention horses bred by humans: do *they* reproduce on their own? What is this "on one's own"? A recursive question in the project of structural anthropology, which regards myth as the endless computation of a loop expressing human origins as either *chthonic* or *autochthonous* (of the earth or as one's own earth): *we came from ourselves* or *we came from others*. This computation expresses the dilemma of the Kantian gap: a thing is itself, but other than itself. Appearances come from things, and they don't.

How did we become Mesopotamians? How did we become formatted as vectors of the agrilogistic program? By attempting to ward off the fear about where our next meal was coming from. This fear is based on an *ontological anxiety*: deep down I know I come from others and am related to others. Logical problems, anthropological problems, ecological problems.

Anxiety is intrinsic to the human, since it's what remains when you subtract all the things onto which it has latched itself, like Alien, to discharge its energy.[29] Anxiety is when things lose their significance, when one is thrown back on oneself, as if knowing oneself as a broken tool that sticks out of oneself, an absurd, disturbing loop. I don't mean that humans are different or unique. Rather, the reverse. Not that bottles of Coke have angst (how do I know? I'm not a bottle of Coke) but that, instead of distinguishing me from other entities (Heidegger), anxiety is how I experience myself as a *thing*. Anxiety shows me that I am an entity among others. And since anxiety is an intrinsic part of human being, trying to rid ourselves of it as agrilogistics promises could only result in violence.

Anxiety is *elemental*. I experience myself as a thing insofar as this thing is no longer objectifiable: it seems to immerse me such that distinctions between self and other, far and near, become inoperative. How Heidegger describes anxiety could indeed describe a zero degree of ecological awareness, a sense of being *a set of things without specific or specifiable members* (we'll clarify that idea soon enough): "neither does anxiety 'see' a definite 'there' and 'over here' from which what is threatening approaches. The fact that what is threatening is *nowhere* characterizes what anxiety is about. . . . But 'nowhere' does not mean nothing. . . . What is threatening cannot come closer from a definite direction within nearness, it is already 'there'—and yet nowhere. It is so near that it is oppressive and takes away one's breath—and yet it is nowhere."[30] The "already-thereness" Heidegger describes so powerfully is a givenness without explicit content, vivid and intense, not blank.[31] The basic anxiety described here is the characteristic attunement of an ecological age in which we know full well that there is no "away"—waste goes *somewhere*, not ontologically "away." Nor is there Nature as opposed to the human world. Ecological awareness is necessarily elemental. Fear of a coming eco-apocalypse covers over the elemental by distorting this threat of nothing(ness) into a fear of something more palpable, approaching from a distance, as if to restore the distance enabled by the now badly broken concept *world*. Elemental anxiety is an existential *Ganzfeld effect*, the term for a visual experience that comes upon one during a blizzard. This effect renders *here* and *there*, *up* and *down*, *foreground* and *background* quite meaningless.

The elemental effect is the inverse of what is called *thing theory*.[32] Thing theory relies on Heidegger's tool analysis. When a tool breaks or malfunctions we notice it. This theory of malfunctioning points out that when things smoothly function, when they just happen, they withdraw from access. When I'm involved in a task the things I involve myself with disappear. Yet the element in which I am involved doesn't disappear. This is a precise definition of the element: the appearance of involvement. It's just that I only experience this appearance obliquely, perhaps as goosebumps or a sense of horror or of bliss.

Paradoxically, the inverse of thing theory is not nothing at all, but what one could call *object theory*. I here use the term *object* in the sense described by object-oriented ontology (OOO, introduced in the First Thread), which argues that the malfunctioning tools we notice depend not only on smoothly functioning without our attention but also on a far deeper being that is strictly inaccessible no matter how deeply we probe or how deeply anything probes—including the "tool" in question. But perhaps *deeper* isn't quite the right term. What we have lost, if anything, is a sense of ourselves exactly as *objects* in this expanded definition. We ward it off or kick it upstairs into the realm of esoteric experience. This has to do with the smooth functioning of the very concept of smooth functioning, an anthropocentric illusion that must be agrilogistic. Hammers and nuclear bombs may function smoothly for me, your average Mesopotamian, but surely not for the lifeforms they affect—including myself, when I get cancer from radiation. Mesopotamia is a vision that things might function smoothly, that malfunctioning might be an accident, a blip, a decoration. It's that idea all the way down to the ontological level, where things are extension lumps decorated with accidents—in other words they are smoothly and uncomplicatedly just what they are.

And, to this extent, we have diverged from the experience of indigenous people. First peoples don't live in smoothly functioning holistic harmony without anxiety; they coexist anxiously in fragile, flawed clusters among other beings such as axes and horses, rain and specters, without a father sky god or god-king. They coexist elementally. Yet because anxiety is still readily available—because agrilogistics has far from eliminated it—the divergence is an unstable, impermanent construct. We glimpse the space of the *arche-lithic*, not some tragically lost Paleolithic. The arche-lithic is a possibility space that flickers continually within, around, beneath, and to the side of the periods we have artificially demarcated as Neolithic and Paleolithic. The distinction of *Neo* versus the *Paleo* is evidence of a whole social and ecological program. Consequently, I shall spell *arche-lithic* in the lower case. It is not a proper name insofar as it doesn't designate something that has *proper* boundaries with distinct and rigidly definable *properties*, let alone *propriety*. The arche-lithic is not the past.

In his magnum opus *Of Grammatology* Jacques Derrida develops the concept of *arche-writing*, from which I derive *arche-lithic*. Derrida's book is a magnificent, strange, and profound exploration of the way the shimmering quality of what he calls writing has been blocked and demonized. "Writing" isn't just scratching marks on surfaces. It's the way a differential play, the tricksy play of nothingness, is in operation everywhere, producing and dissolving distinctions. Such distinctions aren't only epistemological, having to do with language and thought, but also ontological, having to do with what Derrida forcefully calls "flesh and blood."[33]

Arche-writing is the ghostly trace that haunts language, the cascading, spectral play of difference and deferral that makes it work. This working always depends upon some context—literally a con-text, something that goes with the text we are reading, something from which the text can't be so easily distinguished, at least not without violence. This con-text, the beings that go with the text, isn't just cultural or historical, and culture and history aren't just human, as Derrida himself observes. The context is physical. A mark depends on an inscribable surface, just as rhetoric depends on listening or, as Derrida puts it, sound depends on "resonance,"—a readiness and receptivity, an openness that Derrida and Heidegger call *always-already*.[34] Something is always-already "there" for meaning to happen. For there to be a squiggle we count as meaningful, there must be a context of rules about squiggles and meaning and there must also be a piece of paper or a chalkboard or a screen or a wax tablet.

Meaning doesn't happen all by itself. Like me, Derrida is suspicious of the cybernetic (systems theory) excitement about the idea that meaning and "life" can emerge from nothing, hey presto.[35] There is something remarkably ecological about Derrida's suspicion. Writing depends on paper, which depends on trees and water, which depend on sunlight and comets, which depend on . . . if we keep going, we soon discover what I have elsewhere called *the mesh*: a sprawling network of interconnection without center or edge.[36] A haunting ecological vibration already hums within the notion of arche-writing, despite many readers' attempts to put Derrida in a box called *idealism* or *skepticism* or *antirealism*. The term *arche-lithic* only makes this hum a little louder, causing what is already the case to become explicit.

Arche-writing logically precedes the rigid boundary between human and nonhuman. It is "the opening of the first exteriority in general, the enigmatic relationship of the living to its other and of an inside to an outside . . . which must be thought before the opposition of nature and culture, animality and humanity, whether inscribed, or not, in a sensible and spatial element called 'exterior.'"[37] The uneasy nonholistic coexistence evoked here spells trouble for hard boundaries between human and nonhuman, life and nonlife, the Paleo and the Neo—let alone the concept of nature: "The concept of origin or nature is nothing but the myth of addition, of supplementarity annulled by being purely additive. It is the myth of the effacement of the trace, that is to say of an originary différance that is neither absence nor presence, neither negative nor positive."[38]

Agriculture looms uncannily large in *Of Grammatology*.[39] Derrida compares writing with plowing and declares that this comparison is not an accident: "The furrow is the line, as the ploughman traces it: the road—*via rupta*—broken by the ploughshare. The furrow of agriculture, we remind ourselves, opens nature to culture (cultivation). And one also knows that writing is born with agriculture which happens only with sedentarization."[40] Nature, culture, agriculture: the terms are linked historically and philosophically. Derrida writes: "The culture of the alphabet and the appearance of civilized man . . . correspond to the age of the ploughman. And let us not forget that agriculture presupposes industry."[41] That last paradoxical sentence is worth pondering. We like to think that "industry" comes after "agriculture"; heavy machinery is a consequence of agrilogistics. But Derrida's point is that agriculture is already an industry from the beginning, and not just logically but physically: it requires metal, wheels, and all kinds of implements. And it demands an "industrial" view of the world as much as it carves out such a view and literally ploughs ahead with it. It presupposes the "viewfinder" that produces the "worldview," static and picturesque, of stockpiles of stuff in fields and granaries and houses.[42] This is not a coincidence. Writing and the origins of agriculture are deeply intertwined.

Perhaps what we now recognize as writing, done explicitly with wax and stylus and stone and papyrus, develops in the agricultural era to *contain* the implicit shimmer of arche-writing, just as agrilogistics contains the arche-lithic. The containment of the shimmer must constantly risk revealing that shimmer happening not only in what we conventionally recognize as writing, but also in ecological, social, and psychic space. Striving to confine the shimmering to a small region of these spaces, agrilogistics suppresses arche-lithic shuddering, the anxiety of not knowing everything, not knowing the future: the openness of futurality is obscured by planning. To think otherwise than agrilogistically, without (stable, bounded) presence (and the present), must be to think the arche-lithic shimmering around and within the presencing of the agrilogistic.

The arche-lithic haunts the twelve-thousand-year present. The arche-lithic obviates the violence and pride of the Paleo, pride such as the current obsession with the "Paleo diet," a form of ascetic violence against pleasure insofar as it is reflexive, "narcissistic," looping. Violence: like *medieval, Paleolithic,* is an idealized and debased term for a time before the long now. The arche-lithic and its ecognosis are without dichotomies of good and evil, need and want, Nature and Culture, human and nonhuman, life and nonlife, self and nonself, present and absent, something and nothing. I balk at saying *without* in the sense of "utterly without" or "beyond": a Nietzschean formula like that tries to progress once and for all like the modernity of which it is sick.

Since the arche-lithic isn't subject to linear time, it is as much "now" as it was "then," and to assert this is to accord with simple biology: human brains and DNA aren't so different than they were more than twelve thousand years ago. We still walk and sweat and throw. Agrilogistics did not result from some fateful encounter with serpentine knowledge in a loop, another ironic buffer against knowing that *we have never been Neolithic.* And the arche-lithic doesn't simply concern the human. The arche-lithic has to do with a weird logical priority of fuzzy loopy sprawling temporality, and, since it has nothing to do with human versus nonhuman or past versus present, we should be able to find evidence of the arche-lithic outside human "culture." Consider that bacteria *already had genes* that could switch on resistance to a fatal dose of antibiotics: an arche-lithic relationship between the simplest lifeforms and chemicals reduced from future lifeforms or synthesized by humans. A future perfect, time-opening always-already within the very DNA of bacteria everywhere—soil, permafrost, caves, isolated ecosystems.[43] The consequence is that the agrilogistic proliferation of antibiotics in cattle (in far greater quantities than in cattle owners) has pushed bacteria to switch on these genes. The more agrilogistics envelops Earth, the more the arche-lithic seeps in despite it.

The agrilogistic sentence. Agrilogistics is a virus, and what sustains it in human being is viral. A nonhuman code is interfacing with another nonhuman code, an easy-to-replicate pattern that is independent of "me." Something purposeless, something (disturbingly) aesthetic, though the agrilogistic code explicitly bans purposelessness. And therein lies the chemistry of the viral hook. Agrilogistics and the willing that sustains it are paradoxical patterns that deny their patternlike status and deny that this denial is a looplike recursion, another pattern. In one sentence, the agrilogistic loop is this:

This is not (just) a pattern.

There is a fit between *This is not (just) a pattern* and dialetheic ("double-truthed") sentences such as *This sentence is false*. Dialetheias lie and tell the truth at the same time and seem to point outside themselves even as they curl up in a circle.[44] A full investigation of agrilogistic code must therefore explore dialetheias, and we shall do so shortly.

Arche-lithic mind is immersed in a nontotalizable host of patterns that cannot be bounded in advance: lifeforms, ghosts, phantasms, zombies, visions, tricksters, masks. In the welter of patterns outside linear atomic time, it must be the case that at least one pattern is a paradox such as *This is not (just) a pattern*. The agrilogistic virus was not a fatal Edenic apple that showed up one day like a Coke Bottle in a San (Bushman) village.[45] The agrilogistic virus was *co-emergent* with other patterns. Furthermore, because patterns are uncertain, belonging as they do to the realm of appearance, *This is not (just) a pattern* haunts *every* pattern. Is this *just* a pattern? Is there such a thing as a *pure* pattern? Of *what* is this a pattern? It's a pattern—or is it data about something behind the pattern? It is—but it isn't.

The idea that we might be deceived is intrinsic to the agrilogistic virus. The possibility of pretense haunts arche-lithic "cultures" of magic as a structurally necessary component of those cultures: "The real skill of the practitioner [of magic] lies not in skilled concealment but in the skilled revelation of skilled concealment."[46] (I must put "culture" in quotation marks because the term is hopelessly agrilogistic.) Skepticism and faith might not be enemies in every social configuration. In arche-lithic space they might be weirdly intertwined. There is an *ontological* reason why the play of magic involves epistemological panic, giving rise to hermeneutical spirals of belief and disbelief. The dance of concealing and revealing happens because reality as such just does have a magical, flickering aspect. It is as if there is an irreducible, storylike hermeneutical web that plays around and within all things. An irreducible uncertainty, not because things are unreal, but because they *are* real.

The basic pattern is a trickster. The agrilogistic sentence *This is not (just) a pattern* forces its hosts to reproduce it as they try to process the *not* and the *just*. *Agrilogistics is dormant in the arche-lithic*, the continuum of human-nonhuman entailment, waiting for a host to download it. In a loop that fascinates, pretense is pretending not to be pretense. Read one way *This is not (just) a pattern* cancels out its loop form, but on another interpretation the sentence is forever plagued by that loop. Agrilogistics assembles itself from the first interpretation. This is a loop to end all loops. Perhaps it is entirely uncurled, perhaps it is pointing at something. It seems to tie the sentence to a (more) constantly present thing putatively "outside" itself, and in so doing it reduces the sentence to (mere and unnecessary) appearance. Along the same lines the interpretation reduces what is supposedly pointed at to featureless extension. It doesn't matter what your fantasy latches onto, as long as you have one. As if the compulsion of advertising vibrated in the axioms that reduce reality to straight lines. *This isn't just a car, it's an experience*. As if the pattern were not a scintillating appearance but a drab substance that needed to be filled in by some intoxicating fantasy.

A universe of pretense, insofar as it is actual. Patterns that are always recursive. Development of patterns such as the Dreaming, which is what native Australians call the arche-lithic, for them still in process. Or what for the Bushmen such as the !Kung constitutes a primordial world simultaneous with this one, a deeply ambiguous and tricksy "First Order."[47] Development of patterns, not to explain some brute reality but rather as the acknowledgment that reality is not brutish (silent and obvious). Reality as *already-patterned*. I (or history or *Geist* or economic relations or anything else) don't bestow the patterns. Things are not (just) data. Yet one dream thought is the thought that this isn't just a dream. The seduction that the seduction might not be a seduction. The dream of teasing apart dream and reality. Or a meme like this: *The arche-lithic is just a dream*, which is only the inverse of *This dream isn't just a dream*. Such a sentence promises an end to endless hermeneutics, endless anxiety as to the ontological status of things. The process of computing this promise is called agrilogistics. Agrilogistics attempts to erase the intrinsic ambiguity of the arche-lithic. Bushman society has a tolerance for ambiguity that would make the average deconstructor seem uptight by comparison.[48] But, in effect, agrilogistics is also an attempt to erase the ambiguity of its own starter solution, the agrilogistic sentence.

This isn't just a dream. What if the Dreaming itself contained this kind of advert? The Dreaming, a Mystic Writing Pad on which all events are written, provides a map of How Things Are, more than law, more than injunction.[49] The arche-lithic is a welter of code that promises realities even as it withholds them. "Prehistoric" is only the moment at which not enough humans had become susceptible to the agrilogistic sentence to make it appear as if there were no history other than the long history of agrilogistic retreat, known to beginning undergraduates as "civilization." The Anthropocene is simply the moment at which there are enough vectors of the agrilogistic sentence populating enough of the biosphere to exert downward causality on Earth systems.

With respect to Jaynes's basic confusion of "consciousness" and "self-concept," we can thus assert something different. Asserting it will require fewer speculative moving parts than Jaynes does. Hearing voices generates anxiety, which generates more voices, which leads to humans believing in one or two of them, taking them literally: early agrilogistic societies tended to believe that the king was the sole or privileged receiver of the voice(s) of god(s). While Jaynes holds that the modern mind has priority over civilization, it makes more logical sense and requires less cognitive machination to argue that the agrilogistic program has priority.[50] What structures thought is agrilogistics, not the other way around. The voices didn't just show up out of the blue. They had always been there. What changed was our attitude toward them. And this is tantamount to peeling consciousness apart from a self-concept: on the ecognostic view, you can have consciousness without a specific idea of "you."

Believing in the voices as commanding and true stemmed from stress concerning food and the subsequent stress of agrilogistics itself. Agrilogistic success led to the dissipation of voices such that we now consider only two types of people as legitimate voice hearers: the insane and the last few hunter-gatherers. Our thoughts tell us that voices are terribly serious and disturbing, that when you hear them you must be deranged or primitive. We can reverse-engineer this thought with an anthropological insight: *not quite believing the voices* is a hallmark of arche-lithic space because the arche-lithic is the space of the trickster. Complete disbelief would be out of the question because that would be as rigid as belief. We believe in voices far more than first peoples. Not quite believing the voices means violating the Law of the Excluded Middle, a consequence of the Law of Noncontradiction: the idea that there is black and white, yes and no, with nothing in between. Very little in regular experience survives this exclusion. For instance, you can't be "in the doorway" if you believe that you are either inside or outside a room. Motion begins to seem impossible. So many ecological beings are "Excluded Middles" and so much ecological action is in the realm of "not quite" and "slightly," gradations of yes.

Coupling Jaynes's insight with agrilogistics, we obtain this: *the orderliness of "civilization" is and was utterly insane* . . . the product of slavish obedience to a command, just as a computer can't help but execute a program when you hit Return. Jaynes asserts that our very consciousness is in fact a *logical extension* of early agrilogistic mind.[51] As he puts it in his "transition" passages, there are many linking steps between the bicameral mind and what he calls subjectivity—transition points whose boundaries Jaynes sees as thin and rigid (predictably, this gives him agrilogistic logic problems). Even less difficult to grasp is how there's much less time between Jaynes's so-called transition period (between early and modern civilization) and the twenty-first century than there is between 10,000 BCE and the transition period. On this larger timeline it doesn't really matter where you put the transition periods within agrilogistics. It doesn't really matter whether you cleave to Foucault and his epistemes or older talk about the emergence of the "modern self" in the Renaissance. The basic resonance between agrilogistics and all of that eclipses them. What we call modern dualism (the Cartesian self or, even better, the transcendental Kantian subject) is just a variant of the agrilogistic attunement, with its god voices and blind obedience, since the physical body under the dualist paradigm is an android run by the mind.

Jaynes represses the logic of his own argument by "othering" schizophrenia and the bicameral mind. How could the very idea that the gods have departed or that there are no gods (negation) be possible without the continuity of the arche-lithic? To put it another way, isn't the dogged persistence of agrilogistics a symptom of the arche-lithic itself? Agrilogistics is a daft idea, well past its use-by date, yet we just keep on being its vectors. We remain Mesopotamians playing an absurd and destructive game of Whac-A-Mole, habituated to bopping Excluded Middles when they pop up.

Contradictory pests. Recall the first axiom of agrilogistics: *Thou shalt not violate the Law of Noncontradiction*. Yet agrilogistics itself defies the Law of Noncontradiction. The attempt to transcend the web of fate ends up doubling down on it: it *is* the web of fate, the very form of tragedy.

You don't have to cleave to the Law of Noncontradiction to be logical. Consider the case of quantum theory. Young theoretical physicists are acknowledging that the classical-quantum boundary is neither thin nor rigid. Indeed, as Schrödinger observed, if the boundary were thin and rigid, lifeforms would be a paradox. We could scatter into pieces at any moment unless entropy-defying events took place between replicators and cells, on the one hand, and their environments, on the other.[52]

Coherence and entanglement are features of the quantum world that defy our ideas about what things are: they are single, never deviant from themselves; they stay put . . . Coherence is when the parts of an object weirdly overlap so that they become the "same" thing, defying our idea of rigid differences among parts and between parts and wholes. Entanglement is when an object appears so deeply linked with some other object that if the one orients a certain way, the other will *immediately* (defying the speed of light) orient in a complementary way. The objects are separate yet "the same." If we take quantum theory seriously enough, I could be put into coherence or entangled with another entity, although it might be very difficult. After all, these things can happen to objects on scales drastically larger than vanilla Standard Model scales. A tiny but visible (to the naked eye) tuning fork can be put into coherence such that it is vibrating and not vibrating at the same time.[53] A tiny mirror (but much, much larger than a quark) emits infrared in a vacuum close to absolute zero, which is to say, it is shimmering without mechanical input: it is "here" and "there" at the same time. Talk about violating the Law Noncontradiction. Schrödinger's "What Is Life?" is being admired again: if my boundaries weren't fuzzy, if there weren't quantum events occurring everywhere in me, I would be a cloud of powder dispersed at the slightest breeze. *My very solidity* depends upon my being fuzzy.

Like tiny tuning forks and meadows, mathematical sets can be fuzzy yet remain themselves. Cantor's transfinite sets defy the Law of Noncontradiction. Russell machinated mightily with Alfred North Whitehead to tamp down the contradictions in Cantorian set theory, like a farmer trying to get rid of pests. The name for this project against pestiferous nonsense is *Principia Mathematica*. It's a marvelous edifice, gluing math to logic. Hard spadework, but very satisfying: it takes several hundred pages to prove that $1+1=2$. Then along comes Kurt Gödel, who shows that even this magnificent logistical structure can grow weeds. Gödel makes the *Principia* say things like "This sentence cannot be proved." Yet the *Principia* was designed to prove every logical axiom! This means that *in order to function* the *Principia must* be capable of talking nonsense. There is an intrinsic flaw in its logical structure. If there weren't this flaw, it would not exist.

What we are talking about is the very existence of weirdness, and, in particular, what we have been calling *weird weirdness*: the secret link between causality and the aesthetic. "This sentence cannot be proved" is a virulent upgrade of a viral sentence invented by Eubulides in the fourth century BCE: "This sentence is false." The sentence, sometimes called The Liar, is a wonderful example of a strange loop. If *This sentence is false* is true then it must be false, in which case it isn't true. But if *This sentence is false* is false then it is lying, in which case it is telling the truth! *This sentence is false* is a dialetheia, a double-truth. Alfred Tarski invented the concept of metalanguage to deal with this twisty weed of a sentence. If you believe that metalanguages are fine upstanding policemen who always do their job just right, you can invent a rule:

"This sentence is false" is not a sentence.

Now watch as I blow up this concept of metalanguage with a single sentence, which is just a tighter upgrade of the first sentence:

This is not a sentence.

That's much, much "worse," like viruses or bacteria that can become much worse if you try to get rid of them. Agrilogistics and metalanguages are wars against entities seen to contradict the idea that (human) existing is better than anything else.

Like lifeforms and DNA, sentences must contain nonsense in order to exist. And since, along with Laurie Anderson and Edmund Husserl, I hold that ideas and sentences actually are viruses that are mind independent, this isn't just a simile. Sense must coexist with nonsense, its shadow. A thing is shadowed by another thing because it's shadowed by itself. Recursion points to coexistence in a nonholistic, not-all (which is to say *ecological*) possibility space. Metalanguages try to escape this possibility space, to reduce the paradox of coexisting: we *entail* one another and *are not* one another. Plants, specters, and hallucinations return more vividly when you try to prune them. To distinguish thought from nonsense is like taking a lifeform out of its habitat. Deleuze and Guattari put it beautifully: the birth of monsters is hypervigilant reason.[54]

Let's allow arche-lithic nonsense to speak its knowing weirdness. Let's nibble away like field mice at the philosophical axioms within agrilogistics. We'll start by nibbling that rather boring lump of American cheese, the Easy Think Substance. Substances and accidents are not how things are. Things contain an invisible rift that is nowhere in perceptual spacetime: I can think it, but I can't touch it. This is a rift between what they are and how they appear, but since I can't put my finger into the rift, I can't separate these two.

Kant cleaved to the idea that a thing was ultimately a mathematical correlate of itself in a human mind, perhaps in the manner of someone clinging for dear life to a stalk in a flood. Recall that Kant had seen something weird about raindrops: they are what they are yet not as they appear. The stalk Kant clings to says that a thing exists because I can mathematize it. *Mathematics* comes from the Greek *mathēsis*, which means *getting used to*, acclimation. The Tibetan Buddhist for *getting used to* is *göm*, which is also the term for *meditation*. There is mathēsis and there is computation: a limited, logistical application of mathēsis. In the same way meditation consists of awareness, an open part, and mindfulness, a logistical part. Mindfulness can be used as a tool for getting used to awareness, which has a gnostic quality—the strangeness of knowing-in-a-loop about loops. Likewise, because philosophy is not wisdom but the *love of wisdom* (philo-sophia)—always open and strange like awareness—its resting state simply can't be straight or straightforward.

Yet mindfulness might become a way of numbing out and avoiding the strange openness, the love of wisdom. Math and meditation can be very soothing. They might result in openness to phantasms or they might result in seduction by the phantasm of no-phantasm, which is called *reason*. Which would explain why agrilogistics has become Candy Crush at Earth magnitude.

You can become familiar with a stranger (thought, lifeform, stone) such that the strangeness is canceled out. Or you can become acclimated to the strangeness of the stranger. While meditating, do you get used to thoughts in the sense of reducing them to parts of yourself or as "facts of life" that must be accepted? Or the inverse, which is the same: do you get to know them so you can eliminate them as hostile pests? Or do you grow accustomed to their strangeness: their evanescence, their nothingness, their transparency, yet vividness? Do you become aware of minding itself as a stranger? This ambiguity about acclimation deep in the structure of thought is a possibility condition for agrilogistics. Such a hypothesis would satisfy the desire to know how an evidently self-defeating and violent program could continue running—if you like, how human brains got addicted to agrilogistics.

Weird rain. Think back some more to how, buried within the mathematizing thought of Kant, there is a peculiar and often overlooked moment. Kant was nervous about examples, nervous that they would run away from him, and this moment does indeed come in the form of an image: a pattern not totally assimilated into the logical structure, engraved in the space of writing. There are raindrops. You can feel them touching your skin coldly, wet and small. Though these phenomena are not the raindrop, they are inseparable from the raindrop. Raindrops just do feel wet and small and cold to a human.[55] Raindrops aren't gumdrops, I'm afraid. Raindrops are raindroppy: their phenomena are measurably so. But I can't access the actual raindrops. Their phenomena are *not raindrops*. There is a fundamental, irreducible gap between the raindrop phenomenon and the raindrop thing. Moreover, I can't locate this haunting gap anywhere in experiential space or even in scientific space. Unfortunately, raindrops don't come with little dotted lines and a little drawing of scissors saying *Cut Here*, despite philosophers insisting that there is something like a dotted line somewhere on a thing and that their job is to locate it and cut.

Plato compares this cutting to skillful butchery: how to turn an *animal* (already a degraded term) into *meat* (a still more degraded one): a metabolic metaphor for an anthropocentric process. A good philosopher carves the *eidos* at the joints, as if there were dotted lines on an animal telling one which parts were which.[56] Plato exemplifies a pervasive nihilism in Western (that is to say agrilogistic) thought.[57] Plato is a nihilist insofar as he asserts a thin bright line between a realm of false appearances and a realm of realities in the form of reified, constantly present beings. The cut-along-the-dotted-line genre of metaphysics underwrites the dotted lines on diagrams that specify how to turn cows into beef. With the addition of the steam engine, cow appearances could be eliminated entirely in favor of concentrated cow essence, and agrilogistics eventually brought about Oxo, Bovril, and other forms of powdered British cow. The Chicago *disassembly* line gave Henry Ford the idea of massively efficient *assembly* lines. The current obsession with lab-grown meat continues the tradition, and its not-so-well-known employment of fetal bovine serum is yet another form of liquid cow.[58]

Post-Kantian logic tried to close the phenomenon-thing gap. One logical sealant that might do the trick is materialism. Consider John Stuart Mill, a proponent of *psychologism*. Psychologism reduces logic to rules for how a healthy brain functions. Statements such as "If *p*, and if *p* then *q*, then *q*" are, so to speak, percolations of the brain. There is no strange gap between brain and mind. Logic is just an outcome of material reality. Now the trouble is that this statement must also be a brain percolation, which means that there is an infinite regress. One finds oneself incapable of checking what counts as a healthy brain, which is just what one needed to do. It's easier to conclude thoughts are independent of the thinking of them.[59] Schizophrenics are somewhat telling the truth: thoughts are aliens, which is to say they have a family resemblance to hallucinations and to hint darkly at an irreducible *coexistence* of mind and brain where one can't be collapsed into the other. Minds coexist with thoughts that coexist with hallucinations that coexist with brains that coexist with psychedelic plants, arche-lithic beings if ever there were.[60] The intuition appears accurate at a physical level: plant RNA can jump kingdoms, showing up in animals.[61]

When we study the history of reactions to Kant, it becomes peculiarly evident that something like an animism—an awareness of nonhuman agency, consciousness, affect, significance beyond the human—bursts out of his Pandora's box, *in addition to* anthropocentric stories about the human subject, steam engines, and the Anthropocene, with its callous disregard of nonhumans, let alone consumerism with its ravenous desires to eat the world. And that, uncannily, white Western "moderns" have somehow backed into a position not unlike indigenous spiritualities despite and *sometimes ironically because of* our very attempts to leap out of the web of embodiment, indigeneity, dependence on a biosphere, and so on. If this delicious irony is not evidence of the persistence of the arche-lithic outside atomic linear time, it's hard to know what is.

Kant saw the power and independence of that little raindrop and immediately closed the lid on what he saw, just as he was fascinated with animal magnetism, an idea simultaneous with his and Hume's undermining of metaphysical causality theories. Animal magnetism is to all intents and purposes the Force (think *Star Wars*): an all-pervasive energy that causally connects both animate and inanimate objects. We could read the history of modernity as the simultaneous discovery and blocking of nonhuman beings on the inside of psychic, social, and philosophical space, and one of the ways this manifests is in the discovery and simultaneous policing of something like the "paranormal," causalities that do not churn mechanistically underneath things, but that wrap around, flow out of, and otherwise spray and pour out of things, ontologically "in front" of things, not behind them.[62] But this is just basic Hume and Kant "hypostasized"—turned into a physical substance!

Another name for this pervasive force is *aesthetic dimension*. This too has been policed—kept safe from something that looks too much like telepathic influence, though that is strictly what it is if *telepathy* is just passion at a distance.[63] Right now, visualize the Mona Lisa in the Louvre—see what I mean? Something not in your ontic vicinity is exerting causal pressure on you. So the aesthetic and its beauties are policed and purged of the "enthusiastic," buzzy, vibratory (Greek, *enthuein*) energies that shimmer around its fringe, forever turning beauty into something slightly strange, even "disgusting" (at least at the edges) insofar as it can't shake off its material embodiment, shuddery, rich, affective, and effective.

This telepathic Force-like zone of nonhuman energy keeps nuzzling at the edge of modern thought and culture, as if with enough relaxed religious inhibitions and enough enjoyable products humans default to the arche-lithic. Bron Taylor has comprehensively shown how it surges up through the corny cracks in culture, embarrassing us sophisticated academics with its "dark green religion" visible in corporate styling, blockbuster movies, surfing, Disney, and electronic dance music, let alone radical environmentalisms and obscure philosophers.[64] By *dark* Taylor means that humans are capable of committing to an idea that the nonhuman has validity and power and significance all its own, without us. And that we humans, as in *Dark Ecology*'s upgrade of the concept of species, are also on "that side," "without us" in the sense that we perceive we are woven in among the nonhumans in a fabric we didn't manufacture, like seeing ourselves entwined in a deep coral reef while we appear to float on the surface in our myth of human transcendence, looking down at ourselves among the sea anemones and clown fish.

Narcissistic stones. Awareness of ourselves as another "nonhuman" entity has to do with our knowledge, now including logical proofs, that even our thoughts and logical systems evade us. Despite our intentions, they have a life of their own, which means, despite our fantasies that they are totally coherent, they are in fact fragile, like lifeforms. To be a logical system is to be able to speak nonsense because to be a thing is to be nonsensical. Ecognosis has to do with allowing for this nonsensical, pestiferous dimension of things. A thought, a lizard, a spoon veer from themselves. To be a thing is to be a deviation. A thing, a thought, a sentence are per-ver-se. An en-vir-onment is not a closed circle but a veering loop.[65] A thing is in a loop with itself: a thing and a thing-pattern, asymmetrical, which is why there can be patterns at all—which is why there can be replication. Which is why there can be organic chemicals, lifeforms, and sentences about patterns.

Some physicists are now arguing that once patterns are possible lifeforms are not far behind. Some geneticists are now beginning to suspect that life is far more basic to the universe than previously thought. If one extrapolates the data backward, it might be the case that life arose close to the origin of the Milky Way Galaxy and the universe, 13.75 billion years ago.[66] At any rate, the process might well be far less contingent than "dead matter" ontologies allow for.[67] Patterns minimize energy throughput because, as Freud points out, the purpose of life is death. Death—dissipated energy—is what replicants are aiming for in a dance that Arthur Schopenhauer might have recognized. Patterns are how they aim for it. Patterns, in a sense, are the death drive, and beauty is death, a thought common to decadent aesthetes denigrated on all sides for their refusal to buy into "progress" and the yang culture of affirmation and "just do it." Patterns outlast the lifeforms that they give rise to. Stromatolites, fossil records of single-celled organisms, are a case in point. A patterned, bubbly rock found on an alien planet such as Mars *may or may not* be evidence of lifeforms as such.[68] Patterns logically precede life just as they precede truth.

There is something profound and perhaps disturbing about the aesthetic-causal dimension. And about life: "life" is not the opposite of death. The homology between cancer cells and embryo growth bears this out. The only difference is that an embryo becomes shapely through another death process, apoptosis: the dying away of superfluous cells. There is no final resting spot: the pattern is always excessive.[69] Life is an ambiguous spectral "undead" quivering between two types of death: the machination of the death drive and the dissolution of physical objects.

Easy Think Substance theory asserts that lumps of whateverness are *logically prior* to appearances: in order for appearance to happen there must be bland extension lumps from which those appearances can arise. If you put it that way you can see the paradox. Claiming that lumps are prior to patterns is another way of saying that lumps are not excessive. This feeds the materialist belief that lumps of matter were *historically* first. First came atoms and brute nonconscious matter, then sentience showed up. The teleology implicit in this commonly held concept is difficult to accept if you cleave honestly to evolution theory. The idea that sentience or consciousness is some kind of bonus prize for being "complex" or "highly evolved" is suspect from the start because evolution (and chemical reactions in general) is really cheap. The cheapest possible causal pathway will be chosen over others. Congruent with this cheapness is the fact that, as some panpsychist philosophers are now claiming, it is much more *cognitively* efficient to assume that consciousness in some sense goes all the way down.[70]

But what if appearance were inextricable from essence? If such an entwining were thinkable, one could reverse the Marx Brothers joke often cited by Slavoj Žižek, who uses it to argue how existing or being—or whatever that is—is strangely supplementary to appearing: *Chicolini may look like an idiot and act like an idiot, but don't let that fool you—he really is an idiot.*[71] But what if it were also possible to make the joke upside down? *Chicolini may actually be an idiot, but don't let that fool you—he looks like an idiot and acts like an idiot.* If you think that is funny—and that the reversal is funny—you might be ready to allow for appearing to be looped with being in the way dark ecology wants it to be. You may be in touch with the arche-lithic.

What if ecognosis itself were evidence of this intertwining of appearing with being? Ecognosis implies that being and appearing are intertwined because ecognosis bends around on itself. Ecognosis is a self-knowing awareness that doesn't imply an infinite regress of metaness, but a strange loop instead. When they are seen to be happening in someone's psyche, these kinds of level-crossing strange loops are commonly degraded as narcissism.[72] Yet narcissism is always a relating to an otherness: autoaffection is never pure and is always a strange kind of *heteroaffection*, a circulation of energy chasing its tail.[73] What if the phenomenon of self-reflecting in a loop were the equivalent of a not-me, a nonhuman in the very structure of thought?

For many philosophers the trouble with narcissism, a weirdly circular mode in which something coexists with itself, is that it isn't "for" anything. Narcissism is for Hegel the maddening circularity of A=A.[74]

Hegel declared that A=A is outside of logic, yet it was also the possibility condition for logic, the starting position, just as a region of jungle might be seen as a potential feedlot. How could logic even begin, even if we accept that A=A is outside it, unless there were some way to proceed from A=A? But A=A, argues Hegel, is a dead end, the night in which all cows are black.[75] "Proper" logic implies an ontology where things do not circulate into themselves. Reality is taken to be consistent things that never deviate from themselves. Yet to be self-absorbed is to be a thing as such, let alone a thought of a thing. Even in Hegel this is a minimal "thingness," not absolutely nothing at all: after all, there *is* a night; there *are* cows. Objects are relations with themselves logically prior to relations with others. Forks, nebulae, narcissi, and bonobos are narcissists.

The irreducibility of being's circular intertwining with appearing means that dark ecology requires a serious engagement with narcissism, not yet another consumerist-era dismissal or critique or demonization of narcissism. We might even regard such demonizations themselves as ironic symptoms of narcissism. Elizabeth Lunbeck calls it "the narcissism of the theorist," wishing for a world without "needs and attachments."[76] The wish for a cleansed world affects ecological thought. It would be better to start by admitting that one can't escape the narcissistic loop, anticonsumerist diatribes notwithstanding.

Now that we are talking about narcissism, let's talk about narcissi and other flowers. Flowers provide an excellent opportunity to study the arche-lithic thought that being is deeply intertwined with appearing. There is an underground current within correlationism that accounts for nonhuman things in a rather refreshing way, and one of the sources of this current is the Kantian philosopher Schopenhauer. Schopenhauer's thought operates at a nonhuman, vast, disturbing geotemporal scale sufficient for thinking agrilogistics. Schopenhauer argues that plants are manifestations of will: they just grow. He crams his prose about plants with examples of seeds from early agrilogistic times being repotted and sprouting, uncannily alive after millennia, evidence of an undead will.[77] Schopenhauer didn't fabricate this phenomenon.[78]

In this sense plants are like algorithms, since algorithms don't know anything about number: they just execute computations. Algorithms *look and act like* they are calculating. Thus algorithmic models of plants work just like plants, hence the success of the beautiful book *The Algorithmic Beauty of Plants*. To extend our upside-down Marx Brothers joke, a plant isn't just a plant—it looks and acts like a plant too. A flower is a plot of an algorithm. And a trope is an algorithm, a twist of language that emerges as meaning by simply following a recipe such as "Stick two nouns on either side of the verb *to be*." A trope is a flower of rhetoric (*anthos, anthology*). Milton's Satan, a master of rhetoric writhing with tropes, curls around like a snake trying to turn into a vine. He isn't just Satan—he looks and acts like Satan too.[79]

Disturbingly, rhetoric and algorithms and plants and Satan exhibit a degree of intelligence, or not . . . we can't know in advance. Such phenomena present us with the problem of "looking and acting like . . . " Are such performances for real? The trouble is, the irreducible gap between being and appearing—Chicolini acts like an idiot, but he really is one—happens because being and appearing are intertwined—Chicolini really is an idiot, but also looks and acts like one too. Plants haunt us with what Lacan says "constitutes pretense": "in the end, you don't know whether it's pretense or not."[80] They *might* be lying, which in a sense means that they *are* lying. Just as an algorithm could pass a Turing test (I could discern thinking and personhood in its "blind" execution), so plants are posing and passing Turing tests all the time. In looking at a flower, you are doing the flower's job. Bees complete the test all the time by following the flower's nectar lines. Or, as Schopenhauer puts it, plants want to be known because they can't quite know themselves. This subjunctive realm of "might" and "can't quite" is the cognitive space of the arche-lithic, a world of middles so often excluded by infuriated patriarchs.

In the sense that it is in the "looking and acting like . . . " business, flowering is the zero degree of personhood. Nietzsche's Zarathustra memorably proclaimed that people are halfway between plants and ghosts, and this is witty because it is somehow true.[81] Another way of putting it is that the zero degree of being a person could be turned into a sentence that is a strange loop, something like "This is not just a plant." But *This is not just this* is the basic agrilogistic sentence. Remember *This is not (just) a pattern*. A pattern that differs from itself in being itself.

For Descartes, a fundamental uncertainty is key to reasoning that I exist: "Maybe I'm just the puppet of an all-powerful demon."[82] *Maybe I'm not just a person; I look and act like a person—in which case, perhaps I am pretending?* In this sense, paranoia is the default condition of being aware. Before Descartes digs into (Axial Age) theology, whereby a good god would never deceive him like that, he has to traverse a layer of deep uncertainty: "I might be a robot"—to exist is to be paranoid that you might be an algorithm. To be a person is to be worried that you might not be one. We are still in the terrain of Philip K. Dick's *A Scanner Darkly* and its chiasmic investigation of consumerism, Nature and agrilogistics. It is as if Descartes's reasoning recapitulates the ontogeny of "civilization": we begin with archelithic paranoia and try to cover it over with agrilogistic monotheism.

Let's translate the Cartesian creeps into Aristotelian categories of being, animal, vegetable, and mineral. Isn't the paranoia that I might simply be a puppet of some demonic external force just the suspicion that I might be a vegetable? Since we now know about plant sentience, we could acknowledge this to an even greater extent.[83] T. S. Eliot's line about flowers is perfect from the plant's own point of view: "The roses / Had the look of flowers that are looked at."[84] A disturbing chiasmus lies at the heart of correlationism. Or to put it in Schopenhauerian terms, trees and plants want to be known because they cannot know themselves as bodies. "Blind willing" requires "the foreign intelligent individual" to be perceived, to "come . . . into the world of the representation."[85] This is not the neat symmetry of "mountains, Bruce, mountains," but a chiasmus restored to its fully imbalanced, asymmetrical, invaginated form. The two versions of the

Marx Brothers joke are not purely symmetrical. Version 1 (the actual joke) says that being is different from appearing. Version 2 (the archelithic version) says that appearing is indistinguishable from being. Version 1 is about the ontological realm, while version 2 is about the ontic realm, the realm I can point to, the realm of data.

There are traces of Schopenhauer's strange thoughts about plants requiring some Turing test to complete their being in the thought of his predecessor Kant. Kant argues that the decorations and colors of flowers and animals suggest "that their sole purpose is to be beheld from the outside."[86] He says he can't accept this since it would violate a law he likes against "multiplication of principles."[87] Moreover, it risks implying that nature, which for Kant mechanically does things that look nice (like crystallization), was *trying* to look nice. Schopenhauer harps on this in his analysis of plants and ice crystals that *look like* trees and flowers.[88] Since we know about sexual selection—Darwin argues that aesthetic display goes all the way down at least to beetles—the suggestion that Kant rules out is worth pondering.[89]

Kant had it backward: all we need to reach the idea that plants *want* to be looked at is to *remove* a principle or two. The first is anthropocentrism. The second is necessity. Kantian aesthetics depends on a paradox, a purpose of no-purpose. It seems as if a beautiful thing is designed for me to enjoy its having been designed for me to enjoy, and so on. We have a loop: *This is not (just) a sentence with a point*. Such a sentence seems to map onto our agrilogistic sentence. But what constitutes this loop? Consider sexual selection again. Darwin argues that the only reason why I have reddish facial hair and white skin is because someone thought it was sexy a few million years ago, and she probably didn't have a choice. In other words, she wasn't performing something like bourgeois self-fashioning through taste.[90] She cleaved closer even than Kant to the nonconceptuality of the aesthetic dimension. There was *even less purpose*. There's no reason for these huge horns or this iridescent wing pattern. It's actually terribly expensive from DNA's point of view. Just as atelic patterns subtend DNA, so purposive purposelessness depends on weaker purposelessness formats. Under current conditions, DNA wouldn't be able to apply for state funding for its multimedia project; DNA is not unusual in this respect. There is an excess about how the appearance of a thing is always out of phase with its being in such a way that agrilogistic reasoning perceives appearing as decisively separated and superficial.

Since there's no good reason why an insect is gorgeously iridescent apart from the recursive reason that it looks nice, isn't it easy to imagine that the conditions of possibility for human beauty are beautiful flowers, which are also just there to look nice in a sexual display mediated through bees? Sexual activity itself is by no means just heterosexual or monogamous, as trees prove every day by exploding huge clouds of pollen to be spread by insects and birds.[91] Sexual activity is purposeless in that sense. Isn't it possible that the conditions for *that* are to be found *below* plants, in the logical conditions for lifeforms as such—self-replicating loops that are both physical and informational at the same time? There's no good *reason* why squiggles of organic chemicals should "mean" things to other squiggles that constitute their environment. The appearance-thing gap goes down at least to DNA and RNA. And one wonders what causes such things to exist in turn. Isn't it because there is an appearance-thing gap at all as a condition of possibility for existing as such?

This all means that the human-world gap is not the only one. Everything has a gap like that. Correlationism is not false in itself; it is simply the anthropocentrism and the smuggling in of unquestioned metaphysical factoids about substances and extension. Bacteria came before viruses chronologically, but viruses come logically before bacteria. Viruses are the possibility condition for lifeforms: nonliving patterned strands, truly foreign intelligences (to adapt Schopenhauer) that force other patterned strands to go into a loop and become ciphers. The useless beauty of a flower is thus not a cynical ruse to make more plants. It's a viral cipher that serves no purpose, but that, when caught in another system, say a bee's search for nectar, ends up ironically reproducing itself.[92] The human tendency to produce feedback loops at the level of representation is not unique. Viruses, flowers, iridescent wings, Kantian beauty, tropes, earworms, and daft ideas that float around in my head all share something. They are symptoms of an irreducible gap between being and appearing that eats away at the metaphysics of presence from the inside. Viruses and tropes and flowers might not only share some family resemblance. They might *actually* be part of the same physical family.

Flowering is thus indeed a type of "evil," a necessary evil that comes with existing, since existing means having a gap between what you are and how you appear, even to yourself. Flowers of evil. Isn't it the case that the aesthetic dimension has been seen with great disfavor by philosophers who have varying degrees of allergy to the phenomenon-thing gap? Kant shows that this sinister dimension is intrinsic to thinking as such. There are flowers in your head. Kant doesn't make much of a distinction between an actual plant, an arabesque, and calligraphy.[93] There's no way to know in advance what they are for. It's the problem of pure decoration, which is the problem of givenness. Which is, of course, the problem of reason, since reason is just given like a flower that pops up for *no* reason.

Now we are in a position to be able to talk at higher resolution about narcissism, that much maligned state, maligned principally in philosophy by Hegel and Hegelians in denial about the Kantian explosion. This is because Narcissus was indeed a flower: someone stuck in a loop between what he was and how he appeared, even to himself, a loop in which he ended up metamorphosing into what he already was. Narcissus has the look of a flower that is being looked at—by Narcissus. All entities are narcissists insofar as they consist of weird loops of being and appearance. Imagine this Derridean rhetorical flower applying to humans and nonhumans alike: "There is not narcissism and non-narcissism; there are narcissisms that are more or less comprehensive, generous, open, extended. What is called non-narcissism is in general but the economy of a much more welcoming, hospitable narcissism. . . . Without a movement of narcissistic reappropriation, the relation to the other would be absolutely destroyed, it would be destroyed in advance."[94] Narcissism and coexistence intertwine. We want coexistence to mean the end of narcissism, but this is an agrilogistic thought that would destroy in advance the relation to the other. It is difficult to think agrilogistically in the face of our emerging awareness that we are a hyperobject (species) inhabiting another hyperobject (planet Earth). While huge and hard to discern, these "very large finitudes" are such that it becomes obvious how, as Levinas puts it, "'My place in the sun' is the beginning of all usurpation."[95] We know that other humans and other lifeforms are suffering and we also know that their suffering is in part a determinant of our own existence, and at any given moment our suffering is less (because there is one of us and billions of them). Doesn't narcissism in the face of this intuition seem really, really disingenuous?[96]

Yes, if we think that existence means solid, constant, present existence. This belief is based on the fantasy that all the parts of me are me: that if you scoop out a piece of me, it has *Tim Morton* inscribed all over it and within it, just as sticks of English Brighton rock contain a pink word all the way through their deliciously pepperminty tubes. This is not the case. All entities just are what they are, which means that they are never quite as they seem. The first part of that sentence gives us Hegel's dreaded A=A, the night in which all cows are black.[97] For Hegel this isn't even logic yet; it's prior to logic, like a plant or a hallucination. The second part of the sentence shows how being what you are is also a species of loop; as I argued a few pages before, even a night in which all cows are black has cows in it. There are cows in the darkness, cows of darkness. "Equals A" does something to "A," even if that something is hardly different from A. A goes into a loop. It has the look of a flower that is looked at.

By excluding A=A from logic what Hegel is warding off most of all is the possibility that *this loop thinks all by itself*, that it is a kind of artificial intelligence. That it confounds the ability to tell AI and consciousness apart; and, because being and appearing are deeply intertwined, this confounding of noncontradiction goes all the way down with its subjunctive mood and its Excluded Middles. That it was unnecessary to have undergone the linear sequence of logical reasoning, which begins to look like an imperialist leveling of a foreign territory.[98] A logical imperialism reasserts a rigid boundary between the Neolithic and the Paleolithic, suppressing the arche-lithic. What this logical imperialism disavows is the ouroboric, abyssal swirl of A=A.

There is an autoimmunity problem here, having to do with trying to exclude appearance from being, to marginalize it and render it superficial. The more you try to eliminate the Narcissus virus the more you are able to make things sprout flowers, as if things were like the guns the Blue Meanies fire at the end of *Yellow Submarine*: flowers come out of them every time they pull the trigger.[99] Every attempt to reduce a system to simplicity (by firing a gun at it, for instance, or by trying to unzip oneself, like DNA) ends up with the system reproducing itself, flowering into contradiction. For every record player, there is a record called *I Cannot Be Played on This Record Player*. When you try to play it on this record player, it emits sympathetic vibrations that cause the record player to explode.[100] For every logical system there is a Gödel sentence. For every cell there is a virus. For every stem there is a flower. For every lifeform there is death. Try to eliminate the virus and you get a much more virulent one. Autoimmunity is hardwired into the structure of a thing. Or: *a thing is saturated with nothingness*. Entities are so incredibly . . . themselves. Yet in this selfsameness they are weird, self-transcending. The chiasmic, contrary motion of what things are and how they appear makes a mockery of presence. Things emit uniqueness. They *bristle* with specificity. Purple, pale violet, light blue, their soft and sharp spines and flower-spines bristle forth despite me and my subject-object scissions. This flickering between a thing and its appearance is the reason why coexistence can't be holistic. Something is always missing. My self-awareness is a sense of incompletion.

A meadow is a parking lot. A violent nihilism is hardwired into agrilogistics. A spoon could be a potato. A toaster could be an octopus. A meadow could be a parking lot. Hey, let's build one—that sounds like a good idea! Let's build another road on the outer limits of the city! More people, more cars! Eventually this impulse is expressed as happy nihilism, the cheerful manipulation of extensional lumps, manipulation for manipulation's sake. Just for the taste of it. What dark ecology requires is a *nihilism upgrade.* We need to move from the Easy Think Substance to weird substances. We need to go from extension lumps versus accidents versus the absolute void to post-Kantian things suffused with nothingness.

This means that it is now time to inquire about nothingness. Let's begin by considering two basic types of nothing. Start with the default, Easy Think Nothing. Absolute, obvious nothing. After Paul Tillich, let's call this variety *oukontic* nothing.[101] The *ouk* prefix is Greek for "not" or "non." Oukontic nothing means that there is nothing other than substance or things or whatever, just constant presence. There is *not even nothing* other than the universe of constantly present things. You can already see how even this idea contains one of our dreaded loops. *Not even nothing*? Beware, agrilogistic thinkers, of double negations. This sort of nothing deeply resembles plastic substance—it is the flip side of something that is constantly there, remaining the same all the time, just a bland blank. This void is plagued with the same paradoxes and inconsistencies as its cousin, the Easy Think Substance.[102] For instance, since it's absolutely nothing at all, movement across it wouldn't be possible. It can't be demarcated. The most consistent way to think it is with Spinoza: there is substance, and absolutely nothing else, not even nothing.

Consider a set of things. When you remove those things, you have oukontic nothing, unless you allow a set to remain empty. If you are OK with that, you are OK with the idea that there's a subtle difference between a set and its contents, which is absurd from the point of view of Easy Think Nothing. If there is a slight difference between sets and things that sets contain, you can have empty sets that are members of themselves, to Russell's chagrin. A world in which this slight difference is possible is a world in which the species of nothing is *nothingness*, which Tillich calls *meontic* nothing.[103] The *mē* in the term isn't privative in the same way as the *ouk* in *oukontic*: rather than "nonbeing," it means something like "unbeing," "a-being." It's a phenomenon, or is it? You can't quite tell. It causes things to ripple and float and have futurality and dissolve and move. It makes the world go round and it gives you a heart attack. Things literally sparkle with nothingness. They are "alive." Or, rather, "aliveness" is a small region of sparkling that transcends the life-nonlife boundary. Starlight is refracted through the atmosphere and comes onto my retina just so. It twinkles, twinkles.[104] When I see that, I am seeing evidence of a thing I can't quite see called *atmosphere*, thus *biosphere*, thus Earth's magnetic shield. Where does one draw the line, ecologically speaking? Earth weather is influenced by space weather such as solar storms.[105] When I look at the star, I'm seeing a translation of the star in a biosphere-morphic way and in an atmosphere-morphic way and in my anthropomorphic way. Atmospheres and magnetic shields can be as "morphic" as humans can. The star is a sort of unstar when the atmosphere translates it, let alone me. A translation of a poem is and is not that poem.

Schematics of the thing. A thing is a strange loop like a Möbius strip, which in topology is called a *nonorientable surface*. A nonorientable surface lacks an intrinsic back or front, up or down, inside or outside. Yet a Möbius strip is a unique topological object: not a square, not a triangle. Not just a lump of whateverness or a false abstraction from some goop of oneness. When you trace your finger along a Möbius strip, you find yourself weirdly flipping around to another side—which turns out to be the same side. The moment when that happens cannot be detected. The twist is *everywhere* along the strip. Likewise beings are intrinsically twisted into appearance, but the twist can't be located anywhere.

So things are like the ouroboros, the self-swallowing snake. The Norse myth is pertinent: when Jörmungandr, the Midgard Serpent, stops sucking its own tail, that is the beginning of Ragnarok, the apocalyptic battle. Agrilogistics has been a constant process of trying to unloop the loop form of things. Finally to rid of the world of weirdness is impossible, as is devising a metalanguage that would slay self-reference forever. Violent threats can be made: "Anyone who denies the law of non-contradiction should be beaten and burned until he admits that to be beaten is not the same as not to be beaten, and to be burned is not the same as not to be burned" (Ibn Sina).[106] You are either with us or against us. Torture isn't an argument any more than kicking a pebble is, and the threat of torture is no way to display intelligence, let alone proof. The violence of the threat is in proportion to the impossibility of actually ridding the world of contradiction. Beating and burning, something done to cattle and corn, witches and weeds, is not the same as thinking and arguing. Still, in the margins of agrilogistic thought, we cannot but detect the disturbingly soft rustling of the arche-lithic and its serpentine beings. Beings inherently *fragile*, like logical systems that contain necessary flaws, like the hamartia of a tragic hero.

The modern upgrade of the Cadmus myth is the idea of progress, for instance, the idea that we have transcended our material conditions. Harold and the Purple Crayon is a U.S. children's character who can draw whatever he likes with his crayon in the void. Say he is drowning: he can draw a boat. "I am the lizard king, / I can do anything," "I'm the Decider, goo-goo-ga-joob."[107] But if things are nonorientable surfaces, philosophy had better get out of the mastery business and into the allergy medicine business. We need philosophical medicine so as not to have allergic reactions before we mow the allergens down and build a parking lot. To remain in indecision.

Perhaps it would be better to say that the specific allergy medicine we are making is *homoeopathic* rather than allopathic. In other words, and this is an Irigarayan thought (using patriarchal philosophy against itself), one could put binarization into a loop, causing it to become self-referential and thus to violate the Law of Noncontradiction. Rather than fighting violence with an equal violence, one might more successfully undermine its very form with a kind of aikido. Forcing binarization to go into a loop and talk nonsense would successfully open up an exit route from agrilogistics and its Anthropocene.

Ecological awareness is dark, insofar as its essence is unspeakable. It is dark, insofar as illumination leads to a greater sense of entrapment. It is dark, because it compels us to recognize the melancholic wounds that make us up—the shocks and traumas and cataclysms that have made oxygen for our lungs to breathe, lungs out of swim bladders, and crushing, humiliating reason out of human domination of Earth. But it is also dark because it is weird. The more philosophy attunes to ecognosis, the more it makes contact with nonhuman beings, one of which is ecognosis itself. The world it discovers is nonsensical, yet perfectly logical, and that is funny: the sight of something maniacally deviating from itself in a desperate attempt to be itself should remind us of Henri Bergson's definition of what makes us laugh.[108] And this is because, in a sense, to say "Being is suffused with appearing" is the same as saying *being is laughing with appearance.* Ants and eagles cause philosophy to get off its high horse and smile—maybe even laugh. The name of this laughter is ecognosis. You begin to smile with your mouth closed. To close the mouth in Greek is *muein*, whence the term *mystery*, the exact opposite of *mystification*.

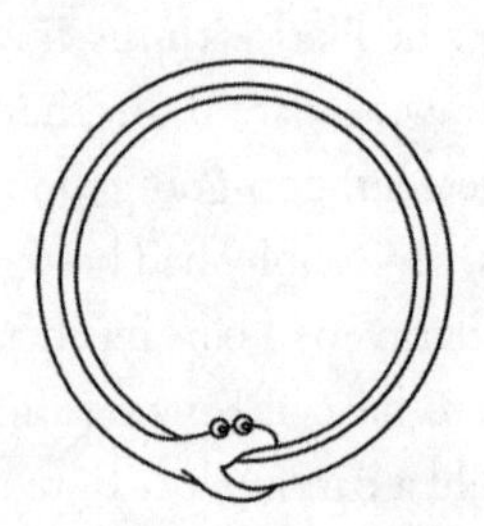

THE THIRD THREAD

Under every grief & pine
Runs a joy with silken twine

What is happening?

I've more memories than if I were a thousand years old.

A big chest of drawers, cluttered with bank statements, poems, love letters, lawsuits, romances, thick locks of hair rolled up in receipts, contains fewer secrets than my sad brain, a pyramid, an immense vault holding more corpses than a paupers' boneyard.—I am a cemetery the moon abhors where, like remorse, long worms crawl across my favorite dead. I am an old boudoir full of faded roses, strewn with a jumble of outmoded fashion, where only plaintive pastels and pale Bouchers breathe the odor of an unstoppered flask.

Nothing's as long as the limping days when, under thick flakes of snowy years, ennui—fruit of bleak incuriosity—takes on immortal proportions.—From now on, O stuff of life, you are mere granite wrapped in vague terror, drowsing in the depth of a fog-hidden Sahara; an old sphinx unknown to a heedless world, forgotten from the map, whose savage mood harmonizes only with the sun's rays setting.[1]

This is sensual Romantic poetry left in the refrigerator too long and blooming with mold: it's one of Baudelaire's *Spleen* poems from *The Flowers of Evil.* Everything slips into an uncanny region, a Spectral Plain on which even the difference between consumerism and ecological awareness flattens out. The breakdown of well-ordered poetry into something like prose is the liquefaction of lineation: writing as plowing neat rows collapses. The narrator tells of being surrounded and permeated by other beings, "natural" and "unnatural" and "supernatural," willy-nilly. The narrator is an abject ecosystem. The Sphinx returns from death, as if Oedipus had failed to kill her off completely. Or as if the spectrality of the nonhuman—can we tell whether they are alive or sentient? Can we tell whether we are?—returns in a Sphinx more fully Sphinx than ever, no longer an Axial Age bogey-being.

Baudelaire seems to have opened a pathway toward a postagricultural ecological age. Some kind of weird punk underground spirit haunts modernity with the specter of nonhumans. Far from being unspeakable within anthropocentric consumerist modernity, nonhumans are showing up all the time, their ghosts leaking into "our" world. We *can* speak ineffable things. When we say we can't speak the ineffable, there we are, speaking it.[2] Secrecy means not totally unspeakable or unknown. You know there is a secret. When philosophy puts its foot in its mouth, its mouth full of itself, tasting slightly different than itself, philosophy starts to smile.

I've been kicked in the biosphere. We live in a reality determined by a one-size-fits-all window of time, a window determined by some humans' attempts to master their anxieties about sustenance. As agrilogistic axiom (3) states, the logistics of this time window imply that existing is better than any quality of existing. So it's always better to have billions of people living near to misery than even millions living in a state of permanent ecstasy. Because of this logic, industrial machines were created. The small rigid time tunnel now engulfs a vast amount of Earth's surface and is directly responsible for much global warming. It's a depressive solution to anxiety: cone your attention down to about a year—maybe five years if you really plan "ahead." One of the most awful things about depression is that your time window collapses to a diameter of a few minutes into the past and a few minutes into the future. Your intellect is literally killing little you by trying to survive. Like a violent allergic reaction or spraying pesticides.

We live in a world of objectified depression. So do all the other lifeforms who didn't ask to be sucked into the gray concrete time tunnel. No wonder then that we find mass extinction depressing and uncanny. We Mesopotamians are shocked to find that our time tunnel is affecting other entities. Evolutionary time and geological time don't work with one-to-five-year time windows. It's only uncanny because we are caught in our viral agricultural app, plowing and plowing away with increasingly strange results, like the winter of 2013 when Minnesota was colder than some parts of Mars.[3]

Let's have more time tunnels of different sizes. Let's not have a one-size-fits-all time tunnel. Let's get a bit playful. Which also means, let's not have a one-size-fits-all politics. We need a politics that includes what appears least political—laughter, the playful, even the silly. We need a multiplicity of different political systems. We need to think of them as toylike: playful and half-broken things that connect humans and nonhumans with one another. We can never get it perfect. There is no final, correct form that isn't a toy. There is no one toy to rule them all. And toys aren't exclusively human or for humans.

If political formations and economic structures are toys, then *oppressive ones are also toys*. Neoliberalism is a toy. Perhaps the problem is not that it is too hard to dismantle—as many a cynical reasoner likes to get paid to point out—but that it is *too easy*, so we miss it. We have perhaps been looking for neoliberalism in the wrong place, as a whole greater than the sum of its parts like an angry Axial Age god. Cynical reason is perhaps a form of monotheism. If we want to coexist ecologically, which is to say animistically and anarchistically, we may need to accept the fact that, while they are physically massive, hyperobjects such as neoliberalism are *ontologically small*, always *less* than the sum of their parts. Since there is only one whole and a multiplicity of parts, and since no object is more real or more significant in some metaphysical sense than another one, the whole must indeed be less than the sum of its parts, however paradoxical we may believe this to be. I tentatively call this characteristic *subscendence*, a once-theological term that describes the incarnation of Christ. The whole is *subscended* by its parts.

Think about this: if you turn 10 percent of your industrially farmed land in Iowa back into prairie, using perennial indigenous species rather than annual cash crops, you achieve a reduction in soil loss of up to 95 percent, nutrient loss reduction of 80 percent to 90 percent, and you reduce water runoff by 44 percent, helping to clean up the Gulf Coast badly affected by agrilogistic chemicals.[4] If you farm rice with ducks, you save countless hours of weeding and generate an ecosystem rich in roaches and small fish, which eat plankton.[5] If forests are integrated with agriculture, agriculture does less damage.[6] Farming and biodiversity are not mutually exclusive. Defying the Law of the Excluded Middle and working with biodiversity at the margins of the farm proves beneficial to all.[7]

Consider another simple modification of agrilogistics. Beetle banks—plowed banks allowed to grow with native grasses that house wintering beetles—shore up biodiversity and in particular the presence of predators who eat aphids hostile to agricultural products.[8] Consider the political action of *copyrighting land as art*. Rather than refusing the category of private property altogether, Peter von Tiesenhausen exploited it. Now the oil corporations that devastated the forests and fields around his house in Alberta, Canada, must find a way not to disturb the surface of his land when they expand.[9] The tactic of copyrighting land is *funny*—it relies on the legalese of loopholes and the playful "uselessness" of art to achieve a tangible effect. This tactic fulfills the wish that politics include the silly rather than exclude it. The ecological reason for this will become quite clear as this thread proceeds.

Look at what the Best Party did in Iceland after the colossal financial crash of 2008. It introduced people, play, and joy back into politics in a serious—but is it serious?—but is it funny?—but is it . . . way. Jón Gnarr, the Mayor of Reykjavik, and his merry band of anarcho-surrealists shook up Icelandic politics and got things back on a good footing through a blend of playfulness and sincerity that is the most congruent ecological affect.[10] The !Kung people among the Bushmen of southern Africa resolve problems through play and laughter.[11] Play is not an accident that happens to otherwise deadly serious utilitarian lifeforms—but why is strict economic cost-benefit analysis associated with being grown-up and "serious"?[12] "If the !Kung visited our offices and factories, they might think we're playing. Why else would we be there?"[13]

Lifeforms play (a cat's nip says "This is a bite and this is not a bite") because play is structural to reality, because things shimmer.[14] A disturbing imbalance and fragility haunts this play in order for it to be play. This is why play isn't just candy or glue but structural to reality. If you think of (agrilogistic) civilization as normative, you have already decided that it is inevitable, and this means that you have decided that agrilogistic retreat is the only way to move across Earth.

Writing at the unsettling start of industrial modernity, Friedrich Schiller was in this sense incorrect to suggest that the (aesthetic) "play drive" could balance the "drives" of "form" and the "sensual" or, in other words, the agrilogistic and Paleolithic. Balance is what play *can't* do: transformative eruption, sure. One detects the difficulty in the way play becomes impossible in the very chiasmus Schiller uses to underline the importance of play, his most famous sentence: "Man only plays when he is in the fullest sense of the word a human being, and he is only fully a human being when he plays."[15] Mountains, Bruce, mountains. If I have to be a full human being before I play—and if I can only achieve this fullness by playing—then play is what I can't do. I am condemned to *wish for play*, a melancholy state that sums up a certain phase of dark ecology. And I am also condemned to a moralistic voice that yells "Play! Or else!" at me like the cans of Coke that tell me to "Enjoy!" or Google that hassles its employees with serious playfulness where what we want is playful seriousness.[16] It sounds like my arche-lithic sensuality is trapped once again in an agrilogistic pipe where play has become a way to make the pipe feel nicer. Play can't simply be recreation, a weekend. That form of play turns out to be a cheap holiday in someone else's misery.[17] The "someone else" being all lifeforms and the "holiday" being agrilogistic built space and its social, psychic, and philosophical affordances.

The trouble resides in Schiller's sense of "fullness" (*in voller Bedeutung des Worts*). To play is to be structurally un-"full," since play is suspended between presence and absence in such a way that one simply isn't constantly present. To be "fully human"—what a drag. We seem to have been trying that for twelve thousand years. Playing as a broken toy among other broken toys sounds more like it. Playing as an arche-lithic state, a trace of the arche-lithic, the beach beneath the street always available by collapsing, flopping, an "uncivilized" deliquescence that also appears too civilized, "decadent." We need our new word *subscendence* to describe it. Subscendence is the inverse of "transcendence," while "immanence" is its opposite. Unlike immanence, subscendence evokes an ontological gap between what a thing is and how it appears, or between a thing and its parts. Play is subscendence, connecting me with the Lego brick, the lichen, the activist network, the microbiome, the melting glacier. We are less than the sum of our parts; multitudes teem in us.

The deliquescent lameness of arche-lithic play counteracts the tensile sporty appearance of play within agrilogistics: the Harold and the Purple Crayon, I can do anything to anything, everything is a urinal waiting for my signature sort of play. Let's not stay stuck in happy nihilism, which is always number one in the charts, the motivation to turn meadows into parking lots because there are no meadows, because nothing really exists. Let's go from happy nihilism to *dark nihilism*. At first dark nihilism is depressing. Then it's mysteriously dark. Then it's dark and sweet like chocolate. You find the sweetness inside the depression. Don't fight it. Find a way to tunnel down. Find a way to see how things sparkle all by themselves. How they play despite intentions. Despite its demonization as narcissistic hippie talk, more and more of us recognize that the personal and the political journeys have the same shape.[18]

It goes like this. We have guilt because we can have shame. We have shame because we can have horror. We have horror because we can have depression. We have depression because we can have sadness. We have sadness because we can have longing. We have longing because we can have joy. Find the joy without pushing away the depression, for depression is accurate. As I noted earlier, one fine day in September 2014 it was announced that 50 percent of animals had vanished in the last forty years. Because of us. I didn't even watch them go. I never personally signed on for this mission. Neither did you. As one of the animals, I never signed on.

Above and beyond the turf wars of academia it's easy to be distressed about the concept of "species" because it's *uncanny*. We notice that we are collectively a zombie just executing an algorithm. It's like the idea of the consumer. I personally never "demanded" products vacuum-sealed in plastic. But I'm told that "the consumer" demanded them. Uncanniness is realizing that something totally intimate is weirdly jutting into your world that was hidden because taboo.

The First Thread stated that "the species thought, stripped of its metaphysical, easy-to-identify, soothingly teleological content, is an uncanny thought that happens not in some universal or infinite realm but at Earth magnitude. It is strictly uncanny in the Freudian sense, if we recall that Freud argues that uncanny feelings in the end involve the repressed intimacy of the mother's body, the uterus and the vagina out of which you came."[19] Freud arrives at this idea of the mother's body by thinking of how the uncanny is excited when the seemingly rigid boundaries between life and nonlife, or between sentience and nonsentience, become confused.[20] Is an android alive? Am I alive? Is an android sentient? Am I sentient? This paranoia is evidence of being physically embodied in an irreducible way. Recall Descartes: being a person means being concerned that you might not be a person. Because there you are, pretheoretically, in Descartes' case sitting by a fire and holding a piece of paper in your hand.[21] And being embodied means *being the biosphere* in the same way as the uterus and the hand turning the key are the biosphere. Being it and not being it.

Ecological awareness is uncanny like that. Freud thinks of being "lost in a forest" and returning again and again to the same spot, like someone in the *Blair Witch Project* or a Cure song.[22] This being lost is already sexualized in a misogynistic way because the example that Freud draws on in the very same paragraph is about being lost in a red light district, trying to exit and coming back again and again to see the prostitutes, described by Freud as somewhere between life and nonlife, puppets in a window who look at you and ask you to buy them.

Realizing we are on Earth in the full Earth magnitude way, realizing that we are permanently, phenomenologically glued to Earth even if we go to Mars, realizing that we are covered and brimful of skin, pollution, stomach bacteria, DNA from other lifeforms, vestigial organs—realizing all this is an experience of the uncanny. Try to strip it away and you are doing exactly what caused the ecological disaster in the first place, trying to come up with one antibacterial soap to rule them all. One extreme symptom would be Nazism, trying to peel the abject embodiment off of oneself once and for all. Or you can become excited about the uncanny. You can get stuck there. You can revel in pointing it out, over and over. What we require is abjection without cleansing, a melancholia without mourning.

You can get stuck in the uncanny because of the prevalence of misogyny. You can juice yourself over and over again with this uncanniness, a popular humanistic sport. It is addicting. If you stay there, it means you aren't really acknowledging the depth of ecological awareness. One's mother's body, the biosphere isn't some abject disgusting thing from which one must distinguish oneself. Underneath the disgust and the horrific uncanny is a type of melancholia, another Freudian term pointing to the indigestible physical and psychic memory trace of other beings within oneself. Indigestible, because once you think you've gotten rid of one, along comes another like the heads of the many-headed hydra. So you can get stuck in tragedy like Wagner and the Nazis, the tragedy of realizing that trying to escape the web of fate *is* the web of fate. Yet within the melancholia is an unconditional sadness. And within the sadness is beauty. And within the beauty is longing. And within the longing is a plasma field of joy. Laughter inside tragedy. Comedy, the possibility space of which tragedy is a rare form. Comedy, the genre of coexistence.

One's mother's body *is* the biosphere. And my stomach that feels like it gets kicked really violently with news of extinction isn't my stomach. I'm not talking about little me, the appearance, suffering here. My stomach is also this biosphere. It implies all the not-me beings.

I've been kicked in the biosphere.

Go right ahead and call it narcissism. Why beat up on a physical good-enough energy feedback between self and environment? Our only task is to include more and more beings within that circuit. It's really just being with the pain without suffering. Let's not stay frozen in horror. Now we know all this information we don't have to keep juicing ourselves. Solutions like geoengineering are ways of not going further, but of being trapped in horror or tragedy. Let's make it down into the sadness and proceed further down from there.

"One rationalization [of the depression epidemic] is fatalism: *I need not act because there's nothing I can do.* False."[23] The caste system is agrilogistic fatalism. You know your place in the highly stratified social body. And cynical-reason pop Foucault now makes television documentaries that say something like: "Our place in society has been allotted to us before we were born." How different is this cool cynicism from Hindu fundamentalism? Human society has been objectively depressive for 12,500 years. Operating within a very narrow temporality tube has been hostile to lifeforms, some of which are humans themselves. It's time to widen the tube.

Tube widening will require extending narcissism. Since we saw that narcissism is an integral feature of being an entity, and that it cannot be destroyed in advance without violence, it follows that extensions of narcissism are one ingredient of future coexistence.

The way in is the way out. We can't get to ecognostic society through further agrilogistic manipulation. We have to step down into things like Baudelaire. Don't try to achieve escape velocity from them, don't try to find some ideal position from where like Archimedes we can leverage Earth, the first image of geoengineering. Let's first step down into a very strange loop indeed, coemergent with the Anthropocene. Down into consumerism. Its histories reduce to a sclerotic and perpetually unexamined meme: *First there was need. Then there was want*. The fact that you can point to almost any period of human history and identify the meme should make us suspicious. But it's also because the logical structure of consumerism reveals the inverse of the meme: *Want is logically prior to need*. It is through this intuition that I am going to argue that there are some ecological chemicals in consumerism, just where we weren't thinking to look. These chemicals are essential for formulating future coexistence.

At first the priority of want seems counterintuitive. Superficially one might claim something like: "You need salt to live." But this is to abuse the word *need*, which evidently has to do with conscious urges. Want, which is desire, is prior to need insofar as desire already transcends my conscious wishes and thus resembles salt and the cellular metabolisms that utilize it far more than it resembles Tim Morton. "You need salt": do you? Well, your cells require it to be cells—but how much? It varies because the homeostatic state of a cell wall changes over time. Metabolism requires varying flows across differing ionic channels, some of which involve sodium. There is no "proper" amount, and the proper is where need lives in historical accounts of want versus need. In those accounts, *need* is precisely calibrated not to be excessive. But this is impossible in a dynamic system such as a single cell. And again a cell isn't you and it certainly isn't conscious you (this is not to say that consciousness is *limited* to me or to humans and so on). *Need* is just the wrong term. It seems the case when we consider that on the cellular level a chemical lack causes all kinds of automated systems to kick in beyond my control. In a perverse way that's much more like desire than need. In the case of salt in particular, it turns out that there is no neurological off switch; your body doesn't care if you have a stroke eating loads more salt than you "need."

Desire is irreducible. To think this drastically upsets the apple-cart of stories about consumerism and stories about how we ruined Earth—Jeremiads worthy of any agrilogistic religion. Environmentalist assaults on consumerism (not to mention Marxist ones, anarchist ones, and so on) paradoxically inhibit subversions of the consumerist possibility space. The space of consumerism—an ultimately artful *ism* of consuming—emerged at the inception of the Anthropocene. With its bohemian consumption for consumption's sake and spiritually enhancing enjoyment (now *there's* a taboo), Romanticism was its quintessential expression.[24] And we are still within that moment of reflexive consumption, just as we are still in the Anthropocene. What is said to be "wrong" with consumerism? It is *for its own sake*. To unthink this wrongness is to think the *arche-lithic*, the timeless time of coexistence. Let's go into more detail about how there reside within consumerism some chemicals that are vital for catalyzing ecological awareness.

This chemical reaction is not an overturning but a *veering under*. Not a rejection of the loop form, correlationism and consumerism, compressed in Lacan's formula for desire, $\$ \diamond a$ and embodied in the advertising and PR tactic of finding a "reason to buy" (a fantasy) and *then* inventing the product onto which this "reason" can latch.

The "reason to buy" is exactly the same as what Lacan means by *a* (*objet petit a*, meaning "little other"). *Objet petit a* is like the carrot on the end of a string tied to a collar around a donkey's forehead or the moon seen from a speeding car: the donkey will never catch up with the carrot and the car will never catch up with the moon. That's in part what the ◇ in the formula means: being a person means being like the donkey or the car, forever chasing a fantasy image that excites desire. The object-cause of desire is not the Coca-Cola or a pink cake in particular: it's the fantasy of why you want to drink a Coke or eat a slice of pink cake in the first place. The ◇ means *is constituted in a loop with*. Tim is a Coke person, not a Pepsi person. The formula $ ◇ *a* is a radical version of Kantian correlationism: "subject" and "object" entail one another in a loop.

In quintessential Romantic consumerism, my object-cause of desire is a certain image of myself. What we are dealing with in thinking consumerism is a confrontation with the loop of narcissism, flickering between autoaffection and heteroaffection. Yet since to be a thing at all, let alone an ecological thing, *is* to be a narcissist, to assault consumerism in a black and white way is to throw the baby out with the bathwater, the baby without which there can be no relation to the other and no possibility of future coexistence at all.

There is something interesting and true about this loop formula for describing what people are. Rather than rejecting this loop form, we simply need to realize that (1) objects of desire aren't blank screens, (2) the loop of desire is a component of a more basic loop between appearing and being at the heart of any entity at all. Point (1) means that a Coca-Cola bottle is not a Pepsi bottle; a frog is not a toaster. In a world where God or some other authority figure is not ordering you to drink Coke rather than Pepsi this implies that all objects appear as they are, unique. Point (2) implies that there is, with all due respect to the anticonsumerism that has dominated ecological thinking for some time, some chemistry in Lacan's Kantian loop that accurately tracks something true about how things are. I know this is counterintuitive, but this means that there must be secret passages from consumerism to ecognosis. Let's try one: *there are looplike entailments between desire and coexistence.*

Ecognosis is abjection. Consumerism is a problem because it unsettles the Mesopotamian idea that we deliberately impose our will on things. The priority of desire suggests we follow directives emanating from thoughts and from Coke bottles rather than deliberately and reasonably "needing" them. There's a fear of passivity colored by a fear of narcissism. This pattern is remarkably similar to the problem of ecognosis. How does ecognosis first appear to itself? As an awareness of things I can't shake off, a *distressing passivity* commonly called abjection. A depressing nausea. The flip side of consumerism expressed in bulimia and anorexia (and punk and Wordsworth and Baudelaire) is abjection, the feeling of being surrounded and penetrated by entities that I can't peel off.[25] There is a path from consumerism to the nausea of coexistence. *Consumerism's flip side is a signal that there are other beings.* Rather than being deliberately conscious of them, I attune to them "passively" since they are already spraying out directives: I *acclimatize* to them (*mathēsis*, *göm*). This drives a huge spike through ideas that my mind is "in" my head and is mine and is the Decider.[26]

Being surrounded and penetrated means that things are always already *given.* I can't reduce this givenness to something expected, predictable, planned, without omitting some vital element of givenness as such. Givenness is therefore always surprising, and surprising in surprising ways: *surprisingly surprising*, we might say. So each time givenness repeats there is no lessening of surprise, which is why givenness is surprisingly surprising. Repetition does not lead to boredom, but rather to an uncanny sense of refreshment. It is as if I am tasting something familiar yet slightly disgusting, as if I were to find, upon putting it to my lips, that my favorite drink had a layer of mold growing on its surface. I am as it were stimulated by the very repetition itself: stimulated by boredom. Another word for this is the familiar Baudelairean term *ennui*. Ennui is the sine qua non of the consumerist experience: I am stimulated by the boredom of being constantly stimulated. In ennui, then, I heighten the Kantian window-shopping of the Bohemian or Romantic consumer.

The experience of vicarious experience—wondering what it would be like to be the kind of person who wears *that* shirt—itself becomes too familiar, slightly disgusting, distasteful. I cannot enjoy it "properly," to wit, I am unable to achieve the familiar aesthetic distance from which to appreciate it as beautiful (or not). Disgust is the flip side of good taste in this respect: good taste is the ability to be appropriately disgusted by things that are in bad taste. I have had too many vicarious thrills, and now I find them slightly disgusting—but not disgusting enough to turn away from them altogether. I enjoy, a little bit, this disgust. This is ennui.

Since in an ecological age there is no appropriate scale on which to judge things (human? microbe? biosphere? DNA?), there can be no pure, unadulterated, totally tasteful beauty. Beauty is always a little bit weird, a little bit disgusting. Beauty always has a slightly nauseous taste of the kitsch about it, kitsch being the slightly (or very) disgusting enjoyment-object of the other, disgusting precisely because it is the other's enjoyment-thing, and thus inexplicable to me. As if I were to find in a junk shop a porcelain vase curiously coated with what turns out, when I bring it close to my face, to be an invisible film of my stomach acid. Moreover, since beauty is already a kind of enjoyment that isn't to do with my ego, and is thus a kind of not-me, beauty is always haunted by its disgusting, spectral double, the kitsch. The kitsch precisely is the other's enjoyment object: how can anyone in their right mind want to buy this snow globe of the Mona Lisa? Yet there they are, hundreds of them, in this tourist shop.

Now in ennui I am not totally turning my back on this sickening world—where would I turn to anyway, since the ecological world is the whole world, three hundred and sixty degrees of it? Rather, ennui is, and this is as it were the Hegelian speculative judgment, the correct ecological attunement! The very consumerism that haunts environmentalism—the consumerism that environmentalism explicitly opposes and indeed finds disgusting—provides the model for how ecological awareness should proceed. A model that is not dependent on "right" or "proper" ecological being, and thus not dependent on a necessarily metaphysical (and thus illegal in our age) pseudo-fact (or facts). Consumerism is the specter of ecology. When thought fully, ecological awareness includes the essence of consumerism, rather than shunning it. Ecological awareness must embrace its specter.

With ennui, I find myself surrounded and indeed penetrated by entities that I can't shake off. When I try to shake one off, another one attaches itself, or I find that another one is already attached, or I find that the very attempt to shake it off makes it tighten the grip of its suckers more strongly. Isn't this just the quintessence of ecological awareness, namely the abject feeling that I am surrounded and penetrated by other entities such as stomach bacteria, parasites, mitochondria—not to mention other humans, lemurs and sea foam? I find it slightly disgusting and yet fascinating. I am "bored" by it in the sense that I find it provocative to include all the beings that I try to ignore in my awareness all the time. Who hasn't become "bored" in this way by ecological discourse? Who really wants to know where their toilet waste goes all the time? And who really wants to know that in a world where we know exactly where it goes, there is no "away" to flush it to absolutely, so that our toilet waste phenomenologically sticks to us, even when we have flushed it?

Isn't ecological awareness fundamentally depressing in precisely this way, insofar as it halts my anthropocentric mania to think myself otherwise than this body and its phenomenological being surrounded and permeated with others, not to mention made up of them? Which is to say, isn't ecological awareness an awareness of specters? One is unsure whether a specter is material or illusory, visible, or invisible. What weighs on Baudelaire is the specter of his bohemian, Romantic consumerism, his Kantian floating, his enjoyment tinged with disgust tinged with enjoyment. The specter of ennui means being enveloped in things, like a mist. Being surrounded by the spectral presence of evacuated enjoyment.

When thinking becomes ecological, the beings it encounters cannot be established in advance as living or nonliving, sentient or nonsentient, real or epiphenomenal. What we encounter instead are spectral beings whose ontological status is uncertain precisely to the extent that we know them in detail as never before. And our experience of these spectral beings is itself spectral, just like ennui. Starting the engine of one's car isn't what it used to be, since one knows one is releasing greenhouse gases. Eating a fish means eating mercury and depleting a fragile ecosystem. Not eating a fish means eating vegetables, which may have relied on pesticides and other harmful agricultural logistics. Because of interconnectedness, it always feels as if there is a piece missing. Something just doesn't add up. We can't get compassion exactly right. Being nice to bunny rabbits means not being nice to bunny rabbit parasites. Giving up in sophisticated boredom is also an oppressive option.

"Gothic" Bohemian decadence is part of a map for future coexistence. Ecognosis is nauseating. Yet an ironic block to thinking this is the truism of modern intellectuals: *If it's depressing it must be true.*[27] Bohemian decadence appreciates a thing for no reason. Consumerism is about (1) having what I think I want, but also (2) following directives emanating from a thing: how to drink that glass of wine.[28] Let's tackle (1), which stems from the irreducibility of desire. Let's return to Lacan's formula for desire, $ ◇ *a*. The "split subject" (little me with my gap between tangible me and the transcendental Subject) is constituted in relation to some object-cause of desire. In plain English, you are what you eat. Or, better, *you are what you desire*. It's a great formula for correlationism: a thing only exists insofar as it's correlated (the ◇ part) with a subject. Correlationism is part of consumerism! And, even better, we can invert the formula so that we get *You eat what you are* or *You desire what you are*. These formulae show that this correlationist consumerism has its nauseous flip side, irreducible intimacy with other beings as a possibility condition for the correlationism and its troublesome extreme—"I can do anything to anything." This possibility condition is not reducible to extreme correlationism. It surpasses it: there are more ways of sinking down into it. This opens a promising line of thought.

People find themselves chasing a desirous reason-to-have or consume. It's like Wile E. Coyote, who chases Roadrunner. Roadrunner is the reason to consume. Wile E. Coyote is little me, the one I can see and touch, with my Möbius twist between me and actual me. Wile E. Coyote will never catch up with Roadrunner. People are constituted as *two* loops, not one: the basic loop of ontic me versus ontological me and the loop between me and my object-cause of desire. Let's call this double loop *Roadrunner*. Roadrunner whispers the possibility that subjects are deeply in-relationship with other entities, even if those entities are fantasy-things. Roadrunner shows there is something alien, something not-you in you. This is promising because we want to allow other beings to exist and we want to care for them without belief. What is called narcissism is the minimal form of this relation to an alien-in-me.

But there is a colossal snag. Lacan was a strongly anthropocentric correlationist, a semi-Hegelian Kantian. His version of Roadrunner is agrilogistic, suffused with sadistic violence. Yet we must sub-vert rather than busting out, because busting out only ever ends up doubling down on what it was trying to escape. We must *subscend*. We must accept the loop of being. We can't cut out our intellect or have an irony-ectomy. John Cleese sues people for acting as if he's Basil Fawlty. In so doing, he *is* Basil Fawlty. What is the case about agrilogistics and its capitalist metastasis is that they are not a solution to anxiety and "need." But this is not because they are inherently evil, which is to say loopy. *It is because they try to smooth out the strange loop*. The twelve-thousand-year trauma of agrilogistics itself, a (flawed) solution to the trauma of hunter-gathering, opens onto the archelithic, our relationship with nonhuman beings: the hot deep rocks in which ancient bacteria persist from some Hadean dawn, inorganic crystal structures, extraterrestrial minerals. A Spectral Plain of ambiguous beings, estuary of the Excluded Middle.

The trouble with consumerism isn't that it sends us into an evil loop of addiction. The trouble is that *consumerism is not nearly pleasurable enough*.[29] The possibility space that enables consumerism contains far more pleasures. Consumerism has a secret side that Marxism is loath to perceive, as Marxism too is caught in the agrilogistic division of need from want. Consumerism is a way of relating to at least one other thing that isn't me. A thing is how I fantasize it. And yet . . . I fantasize, not onto a blank screen, but onto an actually existing thing, and in any case my fantasy itself is an independent thing. This thing eludes my grasp even as it appears clearly. *You are what you eat*. Doesn't the mantra of consumerism (concocted by both Ludwig Feuerbach and Jean Anthelme Brillat-Savarin, almost simultaneously) put identity in a loop?[30] Doesn't this formula hide in plain sight something more than (human) desire? That the "reason to buy" is also a relation to an inaccessible yet appearing entity, to wit, *what you eat*? I imagine what I eat gives me luxury or freedom or knowledge. Yet there I am, eating an apple. *I coexist*.

This can't be! The formula for consumerism *kat' exochēn* is underwritten by ecology! What a fantastic loop *that* is. Consumerism does not twist straight things into a loop. The loop of consumerism expresses and implies beings that are already loops. Sometimes consumerism imagines itself as "I can do anything to anything" sadism, whose artistic double is "Everything is a urinal" constructivism. This imagining and its concomitant social practices are what is violent. Correlationism's hall of mirrors violently represses not a straight real but a looped real.[31]

There's a fundamental ambiguity about consumerism. I love the Coke bottle since I am a Coke drinker. Yet this bottle shows me how to hold it, how to put my lips around it.[32] Anticonsumerism is also Harold and the Purple Crayon. Let's just have a green crayon. Let's make reality once more with feeling in green on a blank slate. Anticonsumerism is consumerism, a mode of performance within consumer space oriented to consumerist objects. It's the idea that I can listen to Coke bottles and crayons that truly disturbs.[33] In this sense consumerism is isomorphic with depression. Depression or melancholia is the trace in one's inner space of what Freud calls *abandoned object cathexes*.[34] The way Freud calls melancholia *a precipitate of abandoned object cathexes* evokes the chemical, preliving structure of the arche-lithic. Depression is the inner footprint of coexistence, a highly sensitive attunement to other beings, a feeling of being sensitized to a plenitude of things. De-pressed by them. So we don't want to reject the logical structure of consumerism. Enjoy a thing just for the taste of it. By listening to it rather than sadistically treating it as silent plastic. Ecognosis means: *letting become more susceptible*.

Melancholy is irreducible *because* it's ecological; there is no way out of abjection because of symbiosis and interdependence. To exist is to coexist. Yet this coexistence is suffused with *pleasure*, pleasure that appears perverse from the standpoint of the subject under the illusion that it has stripped the abjection from itself. Down below abjection, ecological awareness is deeply about pleasure. Ecology is *all* delicious: delicious guilt, delicious shame, delicious melancholy, delicious horror, delicious sadness, delicious longing, delicious joy. It is *ecosexual*.[35] Pleasure and delight only become *more and more* accurately tuned as ecognosis develops.

Anatomy of ecognosis. Ecological awareness is like a chocolate with concentric layers. In the spirit of René Wellek I have mapped these layers in an absurdly New Critical way like some kind of cross between a *Dungeons and Dragons* dungeon master and Northrop Frye. Like Donna Haraway, I believe in the affective power of old-fashioned kitschy theory objects like the Greimasian logic square she dusts off.[36] I'm calling ecological awareness a chocolate in part to provoke the standard reactions: chocolate, sugar, addiction, bad! And to blend that chocolate with ecology (saintly, good, just) in a perverse way.

Each descending layer of the chocolate is a more accurate attunement to the basic anxiety inherent in sentient attunement to things, itself a symptom of the inner inconsistency that marks existence (and coexistence). Machination ruins Earth and its lifeforms, yet it supplies the equipment necessary for human seeing at geotemporal scales sufficient for ecological awareness. We reach for the chocolate because we already attune to the anxiety provoked by this ironic loop of revealing. Something is wrong; our normal machinations (mental and physical) are interrupted or disturbed. We need a piece of chocolate. This is special chocolate, however, that doesn't block anxiety. The basic mode of ecological awareness is anxiety, the feeling that things have lost their seemingly original significance, the feeling that something creepy is happening, close to home. Through anxiety reason itself begins to glimpse what indigenous—that is, preagricultural—societies have known all along: that humans coexist with a host of nonhumans. For reason itself reveals itself to be at least a little bit nonhuman. In turn, reason discovers global warming, the miasma for which humans are responsible. Through reason we find ourselves not floating blissfully in outer space, but caught like Jonah in the whale of a gigantic object, the biosphere. Such an object is not reducible to its members, nor its members to it; it is a set whose members are not strictly coterminous with itself.

The attitude of rigid renunciation in which some ecological speech consists contains agrilogistic code exemplified in the autoimmunity of the Beautiful Soul, an attitude described by Hegel. The Beautiful Soul sees the world as evil and itself as pure.[37] Or the other way around: evil me, pure world (humans are a viral stain). The Beautiful Soul is on an agrilogistic mission, trying to demarcate rigid boundaries. The Beautiful Soul brooks no contradiction: it lives in a world of black and white. It holds an implicit metaphysics of constant presence (the world is all evil, all the time). It's ethically simplistic in a disastrous way. Religious pretensions notwithstanding, the Beautiful Soul is an expression of the worst sort of agrilogistic materialism. Its gaze *is* the very evil that it sees yonder or hither, in the world or in itself. The Beautiful Soul is in a loop that it disavows.

The Beautiful Soul is in tension with the arche-lithic welter of coexistence. It has always been an attunement to ecology. Far from being a form of the subject arising in the Romantic period that is resolved by "true" religion, Beautiful Soul syndrome can be better understood the other way around and as a much more ancient form stemming from a logical priority within thought. Beautiful Soul syndrome is the format of Axial religion with its good and evil, its purities and impurities, its boundary police. And this is because it's already in relation with (other) lifeforms.

The Guilt. Each deeper layer of the chocolate is a phenomenological reduction of the layer around it. Each layer is a certain relation to things, some of which are thoughts. Sometimes it's useful in life, for instance in spiritual practice, to subtract the content of the thought and look at the attitude with which it is being held. This is true in Althusserian Marxism, psychoanalysis and spiritual traditions such as Buddhism. It's not exactly what you think but *how you think it* that poses the problem. From a certain point of view thoughts are viruses that code for specific ways of holding them. So we shall examine not the content of ecological thoughts but the attitude with which those thoughts are held, attitudes that are mutually constitutive of the reality they describe. If we want a good reality—say, for instance, nonviolent coexistence between all beings—we might need to figure out what kinds of attitude are conducive to such a reality.

The outer sugarcoating is guilt. It's addictive. Not that guilt isn't functional for ecognosis. It's just a very low-resolution version of what we discover as we descend into the chocolate. In a sense we *should* feel guilty. Yet there was no sin since loops are not intrinsically evil. Guilt is intimately connected with reification. You have a rigid, crystallized thought about yourself. You try to banish it. This never works. The Guilt is a region of religiose environmentalism. If we stayed here, fundamentalist theists would be right to suspect that ecological information is propaganda that might rob them of their beliefs. It is evident that "lack of scientific knowledge" is a convenient myth that scientism (rather than science per se) tells itself to justify global warming denial, which is far more likely to be based on an *informed* position and thus to involve consciously held beliefs.[38] As Simon Keller puts it, empathizing with global warming skepticism is exactly the tactic we require. Otherwise, and quite rightly, with full respect to scientism's beliefs about belief that cause it to be hamstrung in such situations, deniers will notice that a zero-sum struggle of beliefs is playing out. The mixture of "well we're not 100 percent sure" hesitation and table-thumping scientism at news conferences is deeply frustrating for those of us who know that thoughts are always flavored and colored, never totally abstract and naked.[39] Without validating the deniers, the dominant approach is just preaching to the choir.

The media and the experts often use guilt as a way to force us to be ecological. How's that been working out so far? It's like making us guilty about sugar to force us not to eat it. Guilt is enjoyment upside down. Don't think of a pink elephant! Guilt is a for-its-own-sake that tries not to be so. *This guilt isn't (just) for the taste of it.* Is it superfluous to say that as a product of Axial Age religion guilt is foundationally agrilogistic?

We will find different kinds of laughter on each level of ecological awareness. Laughter here is *guilty laughter*: the uneasy laughter of someone who begins to feel complicit in what they are finding out; the laughter of secret enjoyment. How to scoop out the hidden joy and smear it around a bit?

As we voyage into the chocolate, we'll see that each region has, roughly speaking, an upper and a lower bound. The upper bound of guilt proclaims that you can somehow get rid of the guilt. The lower bound tells you that guilt is irreducible: you will never be able to shake it off. Which brings us to shame.

The Shame. The chocolate layer is shame, just as shame is chronologically prior to guilt in childhood and in human history. It isn't crystallized like guilt; it's a nasty blob of something taken as disgusting, reification experienced as a vague whateverness. Shame does have some ecological functionality. It gets a little bit of a higher-resolution grip on the problem than guilt. This is because shame is deeply connected to being-with: I feel it when I feel others looking at me. Yet I can't stomach the rush to credit shame as the best ecological mode.[40] I feel like killing myself or killing the other when I feel shame.[41] I could feel ashamed of how I've treated coral just by being a member of the hyperobject *species*. But I only have shame if I already *love* coral. It's not good shouting "You should be ashamed of yourself!" The upper bound of The Shame is a violent thrashing whereby I try to rid myself of the stain. Here we find a *shameful laughter* that hides and reveals our deep *physical* complicity with other beings above and beyond the complicity of our enjoyment (guilt).

But the lower bound is just the trace of violence: abjection. Subjects are created when they force themselves to think that they are not made up of abject stuff.[42] They wipe away the abjection, encapsulated in the vomited sour milk on my baby son's pajamas. As in the phrase "Shame *on* you." I can't actually wipe it off. We enter a thick ambiguous boundary between shame and a region that lies within it. At this boundary there is a recognition of trauma, an acknowledgment that we never wiped away the vomit and never could and, by extension, our body, our ancestors in our bones, the fish swim bladders in our lungs, the bacteria in our guts, the phantasms. Isn't this what we encounter daily in ecological awareness? We can't unknow where our toilet waste goes. We think about toxic plastics dripping down our throat when we drink an innocent glass of water. Which is not innocent.

To give us a better feel for the ecology of shame, consider this. Without explicit content, what would the aesthetics of shame feel like? James Turrell is a minimalist sculptor of photons, and his works, such as *The Light Inside,* employ subtle gorgeous electronic light. Turrell is exquisitely attuned to the elemental, which I described earlier as "a givenness without explicit content, vivid and intense, not blank."[43] He exploits the Ganzfeld effect, which, as I described previously, obliterates distinctions between here and there, up and down, foreground and background. One is immersed in vibrant color that seems to come all the way to the tip of one's nose, like rain or cold or tropical humidity. Elemental art: infectious, viscous givenness from which one finds oneself incapable of peeling oneself away.[44] For every subject-object correlation, there is an excluded-included abject.[45] What is skillful about Turrell is his ability to allow us to soothe ourselves into this abjection, as if his Ganzfeld environments were hypoallergenic versions of the things that subtend us.

Disgust has a certain meta-ness to it: it is *the disgust of disgust.* Disgust is disgusted finally with itself, as if it wanted to extricate itself from itself. This dynamic imbalance forms a unit in which—comically from the outside, tragically from within—disgust is unable to leave its orbit around itself. Meta-ness is the very *style* of disgust, meta-ness as sincerity. Despite itself, disgust is elemental. It collapses down into itself, into what we will call The Melancholy. This state realizes that it is abjection liquefied, the meta-disgust melting into disgust. Disgust is "polluted" with itself in such a way that this pollution is discovered to be an irreducible feature of its being.

It's not surprising that modernity, capitalism, and individualism have had trouble with the elemental, seeking to banish it from their easy wipe surfaces. In a society where you are supposed to make yourself, givenness can get in the way.[46] To have spent a lifetime molding oneself, only to find that one's environment was itself our mold, might be disconcerting. The other word for this elemental givenness is *magic*, which is to say influence at a distance: tele-pathy. Elements belong with fairies, selkies, mermaids, traces of the arche-lithic. The arche-lithic is the twisted space of causality itself since the realm of *faerie* is the realm of *fate,* and fate is Urðr, who entwines the causal web. We aren't there yet, but it does appear at this level as if some arche-lithic pixie dust is fogging our dualistic, anthropocentric spectacles. Let's jump further into the fog. It is hard to laugh here, overwhelmed and fascinated by the given. Perhaps just a nervous snicker, like the quiet chuckles as the laughter dies away at the haunting close of Pink Floyd's "Welcome to the Machine."[47] Someone—it seems to be us—has just arrived at a party via an elevator. The room goes quiet. Everyone is looking. Imagine everyone to include nonhumans and there you have it: ecognosis, at least in some form.

The Melancholy. The laughter dies down, and we find ourselves in the space I call The Melancholy. This is the true ecognostic dark night of the soul. By what path do we descend there from the elemental given? It goes like this.

We have been hurt by the things that happened to us. But, in a way, to be a thing at all is to have been hurt. To coexist is to have been wounded. We are scarred with the traces of object cathexes: the very universe itself might be bruised from some unimaginably ancient bubble collision with another universe.[48] To become an I is to risk disavowing these object cathexes, which is what depression is. Deeper into the chocolate we find something softer than guilt or shame. Depression is the imprint of coexistence. Oedipus bears the permanent scar of an intersection of iron, hammer, mountain, and foot, traces of the murderous violence of the father. Yet the very form of our bones is an expression of trauma. Trauma is not only human. The way a glass goblet has been molded is the story of what a glassblower did to a blob of molten glass. The beautiful ridges in the glass are traces of the glass's own lost object cathexes. Things are printed with other things. Something about trauma is nonhuman, and indeed it is the *non-* of any being whatsoever, namely appearance. Appearance is a realm of trauma because appearance is the causal realm, which is to name it the tele-pathic realm. A realm of something slightly "evil," fascination as causal energy. At some point evil starts to smile and appreciate its iridescence. And deeper still it lets go of itself. But it's a long descent until then, a journey with several stages. First we must rappel down what is here called The Horror.[49]

The Horror. The upper bound of The Melancholy is an encounter with horror. At the level of the woundedness there is a reflex, a sudden recoil, a sense that we are too far in and soon won't be able to get out at all. We want out of the chocolate insofar as some concepts that live at the level of horror can be toxic for ecological awareness. We have a problem. We can see it and we can compute it. We are caught in a self-made trap: claustrophobia of the paranoid intellect. Tragedy is the highest form of horror art: we become Oedipus putting his eyes out because he sees clearly, Oedipus from the lineage of weaponized agriculture. Here lives the maniacal laughter of horror. But for all our vivid awareness we are still very much in anthropocentric space: we try to straighten out loops and find the perfect meta position. A sentence in the style of horror writer Thomas Ligotti such as *Human reason is revealed to have been an insect's waking dream* exempts itself from being an insect's waking dream.[50]

In The Horror we encounter the Uncanny Valley. In robotics design it's common to note how the closer an android resembles a human the more frightening it appears.[51] There is an experiential valley where beings such as zombies live in between peaks: we "healthy" humans live on one peak, and all the cuter robots on the other. Zombies live in the uncanny valley because they ironically embody Cartesian dualism: they are animated corpses. They are "reduced" to object status—Easy Think object status, that is—and mixed with other beings—they have been in the soil. The Uncanny Valley concept explains racism and *is itself racist*. Its decisive separation of the "healthy human being" and the cute R2D2-type robot (not to mention Hitler's dog Blondi, of whom he was very fond) opens up a forbidden zone of uncanny beings that reside scandalously in the Excluded Middle region. The distance between R2D2 and the healthy human seems to map quite readily onto how we feel and live the scientistic separation of subject and object, and this dualism always implies its repressed abject, as we have already seen. R2D2 and Blondi are cute because they are decisively different and less powerful. It is this hard separation of things into subjects and objects that gives rise to the uncanny, forbidden Excluded Middle zone of entities who approximate "me"—the source of anti-Semitism to be sure, the endless policing of what counts as a human, the defense of Homo sapiens from the Neanderthals whose DNA we now know is inextricable with our DNA.[52]

As we descend through the abject realm of The Melancholy, we will find that the Uncanny Valley smoothens itself out into a gigantic flat plain. We have already given it a name: *the Spectral Plain*. And we have already encountered it in the poetry of Baudelaire. Ecological awareness takes place on the Spectral Plain, whose distortion, the Uncanny Valley, separates the human and nonhuman worlds in a rigid way that spawns the disavowed region of objects that are also subjects—because that is just what they are in an expanded nonbiopolitical sense. I have called this *animism* in *Dark Ecology* so far, but it would be better to write it with a line through it, as I commonly do: ~~animism~~. A rigid and thin concept of Life is what dark ecology rejects. That concept can only mean one thing: all three axioms of agrilogistics are in play. Life is the ultimate noncontradictory Easy Think Substance that we must have more and more of, for no reason. A future society in which being ecological became a mode of violence still more horrifying than the neoliberalism that now dominates

Earth would consist of a vigorous insistence on Life and related categories such as health. It would make the current control society (as Foucault calls it) look like an anarchist picnic.[53] If that is what future coexistence means, beam me up Scotty. The widescreen view of dark ecology sees lifeforms as specters in a charnel ground in which Life is a narrow metaphysical concrete pipe. Death is the fact that ecological thought must encounter to stay soft, to stay weird.

In ecological awareness differences between R2D2-like beings and humans become far less pronounced; everything gains a haunting, spectral quality. This is equivalent to realizing that abjection isn't something you can peel off yourself. The Nazi tactic of peeling off abjection while supporting animal rights isn't inconsistent at all. Consistency is its very goal. Nazis are trying to maintain the normative subject-object dualism in which I can recognize myself as decisively different from a nonhuman or, to be more blunt, a non-German, a recognition in which everything else appears as equipment for my *Lebensraum* project. So there is little point in denigrating ecological politics as fascist.[54] But there is every point in naming some Nature-based politics as fascist. Here is a strong sense in which ecology is without Nature.

Scientistic speculative realism lives in The Horror, the top level of the realm of abjection, the level where we have not yet discovered the Spectral Plain. Dallying on the slopes of the Uncanny Valley, what is unthinkable and unspeakable becomes the favorite philosophical thought, and the favorite attitude with which this thought replicates is horror—a demonic version of the wonder that Socrates and Plato imagine to be the basic philosophical affect. Perhaps monsters violate the Law of Noncontradiction and perhaps this is disturbing, demonstrating the impossibility of Life as a concept.[55] So speculative horror eliminates the concept Life altogether to abolish the loop implied in the contradiction. Unable to rest in contradiction, horror wraps its terrified hands around the contradiction with bleak certainty.[56] This is yet another paradox of being in a loop while disowning being in a loop. Clinging to the spectral as the horrible is not to have fully relinquished Life, which, as we have seen, is toxic in its Easy Think sense. What I defined earlier as "life" (small letter *l*) is a quivering between two kinds of death. A being quivers and the emotion that flavors that being's phenomenon in me is quivering. Quivering, not creeping flesh that I want to rip off me. The reaction is from the realm of the rubbernecker, even if they are masochistically rubbernecking their own ontic or ontological demise.

A war of escalation arises in which bigger, badder, scarier versions of the thing in itself spawn. What is the phenomenological chemistry of horror? Recall that this is a horror that the nonhuman exists in some strong sense. Something is noncorrelated, but I can think it. The flip side of a being outside thought is the thought that paradoxically becomes aware of this being. The flavor of that thought is a sentence like *This isn't (just) thought in a loop*. I'm caught in Alien's web like the fascinated doctors in *Alien Resurrection*.[57] And I'm loving it, albeit in a suicidal way. A masochistic machismo reigns, according to which I prove that my upside-down Satanic version of an Axial Age monotheistic god (perhaps it's called Cthulhu) who wants to kill me is much more horrific than yours.

We have seen how ecological awareness is depression and how agrilogistics is depression. Why the ambiguity? Depression can lead to an autoimmune syndrome just like an allergic reaction, a reaction not unrelated to agrilogistics. Depression is an allergic reaction of the intellect to its host, the poor sentient being, embodied and fragile, akin to how white blood cells start to attack the body that spawned them. They go viral, cleansing the world of ghosts and spirits, the "pathetic" sensations and feelings.[58] The extreme political variant is accelerationism: capitalism should be sped up in the name of anticapitalism to bring out its contradictions, with the hope (underline hope) that it might then collapse. But this bringing out is thought as inviting a colossal machinelike alien from the future to come and destroy us pathetic humans once and for all. To allow agrilogistics to destroy its host: the future, minus Earth. That is what accelerationism is hoping for. The name of this hope is despair.

Intellect is indeed an organ of extinction and that is not great.[59] You can't snicker triumphantly about that. Like all autoimmune systems, intellectuality-depression reinforces itself. How does one sub-vert a self-reinforcing loop? To cast intellect away would be the absurd anti-intellectualism that is part of the problem, trying to return to some state of Nature defined by stripping "civilization" of its symbionts: intellect, plastic, cancer; and, beyond this, stripping the very loop form that provides the structure for beings. This is absurd because "civilization" is already agrilogistic scorched earth and retreat with the nonhumans airbrushed out. Stripping the human realm of its symbionts is . . . agrilogistics. The deadly seriousness of Justine in Lars von Trier's *Melancholia* is evocative of speculative realist horror. "I know things. And when I say we're alone . . .we're alone [in the universe]."[60] *We're alone* means that there is just dead matter, and the accidental sprinkles of sentience and intelligence are just a highly contingent blip in a tiny meaningless part of the darkness.[61] This isn't scientifically accurate, though it claims to be. And it accepts profoundly the Easy Think Substance: first there were lumps of whatever, then there were pretty, mostly pathetic appearances. We have become allergic to chocolate. But we are too far in. We can't make our way back up to good old guilt. We have too much ecognosis. We need to find an alternative to horror as a host for intellect. Not aside from intellect, but inside. A homeopathic remedy.

We need to find within horror some form of *laughter*. Let's start then with something funny about intellect itself. The face of horror-knowledge is nothing but the face of the boy (underline *boy*) Macaulay Culkin in *Home Alone*. The stereotyped behavior of someone locked into their style without knowing is inherently funny.[62] Laughter becomes *ridicule*. And without quite realizing it at first, we will have discovered that at this stage in our descent into dark ecology we seem to have entered a region called the *Realm of Toys*.

The Realm of Toys. And the first toy is the style of horror! We'll shortly discover that the Realm of Toys provides the blueprint for an ecological polity, a polity that includes nonhumans as well as humans. Attempts to fashion a politics at the upper layers of ecognosis risk violence. It's strange and fitting that less violent future coexistence should be found within the deeper layers. Perhaps this is why it's hard to articulate. An ecological politics based on guilt underlies "return to Nature" tactics. Basing politics on horror necessitates some kind of resignation tinged with Schadenfreude, wide-eyed and screaming while Rome burns. What mainly impedes ecognosis is the deadly seriousness; we require toy political forms that don't take themselves quite as seriously. A lifeform, an engineering solution, a social policy, another lifeform—they join together and become another type of toy, in a sort of ecological Lego. Not as models for some future serious form. Why? Because of interdependence, there's always a missing piece of the jigsaw puzzle. There can never be The Toy, one toy to rule them all. Attempts to impose one system top down have consistently failed to feed as many people as a variety of smaller scale approaches.[63]

The "Crash on Demand" thought of David Holmgren resonates with the view I outline here. For ontological reasons if not for political ones, top-down solutions are strictly impossible, such that (doomed) attempts at their imposition must be violent.[64] The name for this violence, in a world of shrinking oil reserves and nation-states, is *brown tech*. This is why Hermann Scheer advocated that small communities such as towns get hold of their own energy supply, delinking from the oil-based energy grid and its top-down, one-size-fits all violence.[65] Even the UN is now admitting that small-scale organic farming is the only way to go.[66]

Play versus being caught in the headlights. Stripped of reification, horror is more like what the Latin *horrere* means—just "bristling," the hairs on your body standing to attention. Theodor Adorno says it best:

> Ultimately, aesthetic comportment is to be defined as the capacity to shudder, as if goose bumps were the first aesthetic image. What later came to be called subjectivity, freeing itself from the blind anxiety of the shudder, is at the same time the shudder's own development; life in the subject is nothing but what shudders, the reaction to the total spell that transcends the spell. Consciousness without shudder is reified consciousness. That shudder in which subjectivity stirs without yet being subjectivity is the act of being touched by the other. Aesthetic comportment assimilates itself to that other rather than subordinating it. Such a constitutive relation of the subject to objectivity in aesthetic comportment joins eros and knowledge.[67]

"That shudder in which subjectivity stirs without yet being subjectivity is the act of being touched by the other." In other words, this horror format is ecological, an aspect of the feeling of being alive. These are the very last sentences of a very searching book: Adorno really *really* means it, and Adorno is a philosopher who only really means it.

Making toys would include meaningful collaborations between the arts, the humanities, and engineering, rather than the mutual suspicion that reigns today. Humanists are hamstrung by cynicism and engineers are hamstrung by research contracts from big corporations. But we all make toys—toy worlds, prototypes, forms to think with, in our heads, on paper, in wood and plastic. What if humanists worked with engineers on suggestions and models for far future toys, toys not bound by the temporality of current corporate needs? One could take one's prototype pacemaker and redesign it for a future two thousand years from now when the climate had changed by these and those factors, where different lifeforms were dominant. Part of the benefit of this practice would be to create not only interesting toys but also a much needed host of interesting *toy temporalities* that reach beyond the agrilogistic temporal diameter.

Toys are suspended between being and appearing. *Toy* is an umbrella term for anything at all.[68] Toys play. There are things, and that's why we have appearance. But we can never peel the appearance aspect away from the thing. When a cat nips you, she is saying: *This is a bite and this is not a bite*.[69] She is contradicting herself yet telling the truth. Nonhumans know how to play. A thing plays in order to be a thing. Heraclitus: nature is a child at play. You don't have to be frozen in horror at the ghosts even if you can't get rid of all of them.

When they ran Reykjavik, the Best Party toyed with the idea of not having a one-size-fits-all top-down final approach to politics. Perhaps the Best Party is a model of future ecological political forms. Future coexistence—from a fully future future, not a predictable extension of agrilogistics—accepts contradiction and ambiguity. It is neither a progression nor a regression from contemporary consumerist agrilogistics. It has accepted the fact that we Mesopotamians never killed the loopy Hydra. Humans should act to change their material conditions, but those conditions aren't an Easy Think Substance. Those conditions might be wasps, mycelium, spores, and leopards. We are lazily used to our ontology coming with an easy to discern, snap-on ethics or politics and vice versa, rather than as complex Legos we have to assemble. Consider how we might recover from agrilogistics. The point would not be to dismantle global agriculture and replace it with yet another top-down solution. Instead we need many toy structures, many temporalities.

Shelley wrote that humans lack the creativity sufficient to imagine what they know, and that was in 1821. By contrast, conventional Marxism and scientism imagine that nothing is wanting in our idea of knowledge. The problem is we haven't been pure enough, strong enough, or committed enough to our beliefs. But if our "calculations have outrun our conceptions" then to act within the well-worn grooves simply prolongs the problem.[70] Toys connect humans with nonhumans: a child's hand with a robot's arm, a piece of lettuce with a rabbit. And toys are nonhumans in themselves. Because of the nature of ecognosis, an ecological future is toys at every scale without a top level that makes everything sensible, once and for all. Perhaps that was the problem all along. Suddenly, horror appears ridiculous and another kind of laughter breaks out.

The Ridiculous. In the middle of John Carpenter's movie *The Thing*, when it couldn't get much worse, the viral, morphing, oozing alien who imitates others, tricking and then devouring-imitating those who interact with it, has absorbed one of the characters in the Antarctic research station. The remaining crew are busy blowtorching most of the dripping Thing. But some of The Thing's mass escapes their attention, in the form of the head of its latest victim. Under the table the upside-down head sprouts spiderlike legs and begins to crawl out of the door, emitting weird breathlike distorted moans. It is at this point that one of the crew utters the immortal words "You gotta be fucking kidding," upon which they torch the spider head.[71] As we might surmise from the image of torching a toy that looks like a human-spider hybrid, ridicule can go either way: toward violent destruction of the abject or toward acceptance. Acceptance is achieved by turning the laughter back around on the horrified subject.

In this region of The Melancholy, we encounter the art of the absurd. The gallows humor evoked by the plays of Samuel Beckett is poised between laughter and despair. In this subregion, toys appear to be demonic puppets. There are still human masters who differentiate themselves from the puppets. There is a *toying around* at this level, a mistreatment. The Ridiculous is a place where toys are torched for being anomalous. Large patches of the Law of Noncontradiction still predominate at this level. The Ridiculous is a realm of satire and sarcasm, comedy with something missing. A meta-ness lingers here.

In The Horror the ecological emergency looks like tragedy. But, as the ancient Greeks knew, the tragic can be viewed another way to bring out its implicit comedy. For every three tragedies in the Athenian City Dionysia there was a satyr play followed by two days of comic poetry and drama. The satyr play grounded the agrilogistic machinations of tragedy in its ambiguous arche-lithic boundary state between the Neolithic and the Paleolithic where "monsters" and human-nonhuman hybrids (satyrs and centaurs) roam. Aeschylus' Theban plays were followed by a satyr play called *The Sphinx*, that agrilogistic boundary-being. That *The Sphinx* should follow the one who forced the Sphinx to kill herself is a wonderful enactment of the way the arche-lithic keeps growing out of the cracks in the concrete. The satyr play was ribald, physical, bawdy—something like a gross-out comedy that relieved the tragic tension, placing it within a larger context of ridicule. The very attempt to escape the web of fate *is* the web of fate; then you notice that you are caught in the loop and you are the loop. Two days of laughter follow. It is as if by marking the boundary of agrilogistics the satyr play summons something from the equiprimordial space of the arche-lithic, the ambiguous realm of coexistence between humans and nonhumans.

The Ethereal. At some point you stop wanting to apply flames to contradictory toys. You start to collect them. A less violent abjection broods here like a pale mist. It is almost beautiful. We discover the whimsical toys of kitsch. The Ethereal is suffused with a strange goth feeling, like the room of replicant designer J. F. Sebastian in *Blade Runner*.[72] *Goth* here means the abject (and highly popular) underside of Romanticism, slightly too melodramatic and dark, and inclusive of pleasure—but a weird pleasure. Not pushing things as far as horror, which would reproduce the basic phenomenology of the subject-object dualism even without explicitly clinging to it as a theme. Less Robert Frost than Robert Smith, goth suspends the experiencer/

consumer in an Excluded Middle state of *slightness*, slightly twisted eroticism, slightly dark but not overloading the system with horror and thus forcing it to give in and be pulled up toward the shallower modalities of ecological awareness. Baudelaire intrigued by his abjection, sitting alone and feeling weird without recoiling in horror and without contextualizing his experience, as if beauty were still possible, but only on condition that we drop the anthropocentrism and relate to a truly unconditional beauty, including the unconditionality of no (human) standard of taste—the fringing of beauty with fascination, disgust, and fear without trying to airbrush them out of the picture. Disgust, fear, fascination are evidence of not-me entities attached to me or rapidly approaching.

Kitsch is others' (inevitably weird or disgusting) enjoyment objects, evoking the intrinsically nonhuman aspect of enjoyment as such. "How on Earth could so many people want a snow globe of Gandalf?" But this gives rise to a very valuable insight. Even when I am having it, enjoyment isn't "mine." It exceeds my conceptual grasp. I'm sorry to break it to the avant-garde, but art, in an ecological age, will melt into kitsch because there will be no single, authoritative scale from which to judge. To accommodate this will require a major revaluation of kitsch, salvaging it from its trashy category as a trash can for things we think of as fascist or otherwise beyond the pale.[73] The pretheoretical rejection of symbolism in theory class will need to explain itself without resorting to knee-jerk disgust. The unshocking idea that art should shock the bourgeoisie out of its complacency is what needs to be gently folded and put away. Again, the chemistry of consumerism should not be rejected, because it is consumerism that opens the world of kitsch.

As we descend into The Melancholy. the difference some want to maintain between *interest* and *fascination* evaporates as the not-me object exerts its gravitational pull. The guardian of this region is Wall•E, the garbage-collecting robot who maintains a collection of gadgets and trinkets the humans have left behind on a trashed uninhabitable Earth.[74] The exit from modernity is somewhere in the craft store with its gaudy, sparkly Christmas ornaments for trees full of light, pagan consumerism. There are no longer piles of trash because there is no longer anthropocentrism. What seemed to be trash are objects with what Jane Bennett calls a vibrant life of their own.[75] A whimsical laughter resounds, the laughter of Miranda from Shakespeare's *Tempest* marveling at the strangeness of other lifeforms. Fascinated, I begin to laugh with nonhumans, rather than at them (horror and ridicule), or at and with my fellow humans about them (shame and guilt).

It's Wall•E's automated melancholia that saves Earth, an affect liberated from its human casing, relentless yet soft and sad. Wall•E is the guardian, not Justine of Lars von Trier's *Melancholia*. Justine is a guardian of The Horror. She (thinks she) knows too much. Wall•E doesn't have a clue. Precisely because of this, he doesn't stand for the absolute abandonment of hope. Without horror, depression *is* a collection of other entities for no reason, a pile of garbage mournfully, lovingly preserved as in a museum or a zoo. This is a remarkable place, this sad aestheticism without standards, this collection of kitsch. But still one awaits something better. Frozen mourning and obsessive cleaning do not exhaust ecognosis. The Melancholy doesn't know what the toys want. Yet something very significant has occurred: progress down through The Melancholy has seen the Uncanny Valley flatten out into the Spectral Plain. There are no subjects and (non-OOO) objects anymore, just various kinds of specter. If you are going to be the specter of communism, you had better talk from this level (or lower) and stop being haunted by the specter of the nonhuman.

The Hollow. Something strangely beautiful lies in the region below. We can detect it in The Melancholy. It is as if sparkles and shards of what lies below make their way upward, twinkling unconditionally in the fascinating reflections of the mournful toys. But to get to that strange beauty we must swim down through The Hollow, a boundary region. Trying to escape depression is depressing. We begin to recognize this loop as a hollowing out. We begin to learn from the no-way-out-ness of Alice, trying to leave the Looking Glass House and ending up back at the front door.[76] The hollowing of depression, in turn, is recognized as a *thing*, which is to say a thing in all its withdrawn mystery.

What is happening as we enter this chocolate? A perverse inverted Hegelianism. Not Hegel upside down in the Marxist sense, nor what Hegel calls *Aufhebung*, where we keep transcending and transcending further and further out. We are collapsing down into a throng of more and more real objects: subscendence. By *real* I mean not reified, not depending on a subject, not undermined or overmined: not reduced to atoms, fluxes, or processes or reduced upward to correlates of some Decider. A weird joke is in process. Perhaps its collapsed style is best caught by Syd Barrett, inventor of glam and goth and whimsical toys, out of his mind and depressed and sad and The Piper at the Gates of Dawn:

> And the sea isn't green
> And I love the Queen
> And what exactly is a dream
> And what exactly is a joke?[77]

Now you see me, now you don't. Fleeting laughter resounds. We begin to enjoy contradiction ("And the sea isn't green"). We begin to relax our defense against ontological paranoia ("And what exactly is a dream"). We relish in ambiguity ("what exactly is a joke?").

The Sadness. Inside the congealed Hollow is a liquid Sadness. This sadness is not the trauma of relating to one's wounds from other things, the wounds that make me what I am. This sadness is a liquid inside the wounds. It does not have an object; it *is* an object. This being-an-object is intimately related with the Kantian beauty experience, wherein I find experiential evidence without metaphysical positing that at least one other being exists. The Sadness is the attunement of coexistence stripped of its conceptual content. There is a sad laughter of coexisting, beginning to believe in its magical powers, like the poignant recognition-misrecognition of the cross-dressing characters in Shakespeare's *Twelfth Night*.[78] Since the rigid anthropocentric standard of taste with its refined distances has collapsed, it becomes at this level impossible to rebuild the distinction we lost in The Ethereal between being *interested* or *concerned with* (this painting, this polar bear) and being *fascinated by* . . . Being interested means I am in charge. Being fascinated means that something else is. The fascination of beauty is what some philosophy tries to ward off at all costs.

When you experience beauty, you experience evidence in your inner space that at least one thing that isn't you exists. An evanescent footprint in your inner space—you don't need to prove that things are real by hitting them or eating them. A nonviolent coexisting without coercion. The basic issue with beauty is that it is ungraspable. I can't point directly to it and I can't decide whether it's me or the thing that is emanating beauty. There is an undecidability (not total indeterminacy) between two entities—me and not-me, the thing. There is a profound ambiguity. Beauty is sad because it is ungraspable; there is an elegiac quality to it. When grasped, it withdraws, like putting my hand into water. Yet it appears. This thing I am finding beautiful is beautiful to me. It is as if it is definitely this thing and not that thing. I have accepted that a thing is a narcissist; I have stopped trying to delete my own narcissism. The beauty experience just is narcissism, inclusive of one or more other entities. A narcissism in me that isn't me, including me and the thing in its circuit: ecognosis.

Beauty is virtual: I am unable to tell whether the beauty resides in me or in the thing—it is *as if* it were in the thing, but impossible to pin down there. The subjunctive, floating "as if" virtual reality of beauty is a little queasy—it's captured well by Nicholas Royle's use of the term *veer* to describe aesthetic experience. From *veer* we obtain *environment* and *perversion*.[79] When a ship is veering, it's not certain whether it's acting on the ocean or letting the ocean act on it. In the same way, beauty requires a veering toward a thing. The thing emits a tractor beam in whose vortex I find myself; I veer toward it.

The aesthetic dimension says something true about causality in a modern age: I can't tell for sure what the causes and effects are without resorting to illegal metaphysical moves.[80] Things influence one another such that they become entangled and smear together. Something slightly sinister is afoot—there is a basic entanglement such that I can't tell who or what started it. Beauty is like a human-flavored, massively magnified version of what happens to a tiny tuning fork when you put it into a vacuum close to absolute zero. It starts to show you that it is—and isn't—there all at once: in the haunting phrasing of Aaron O'Connell, who ran one of the experiments, it "breathes."[81]

Beauty is the givenness of data. A thing impinges on me before I can contain it or use it or think it. This impingement is not susceptible to being pinned down. It is as if I hear the thing breathing right next to me. And that is the true origin of the uncanny inertia we sense in its proximity. Something slightly "evil" is happening: something already has a grip on us, and this is demonic insofar as it is "from elsewhere." This "saturated" demonic proximity is the essential ingredient of ecological being and ecological awareness, not some Nature over yonder.[82]

Interdependence, which is ecology, is sad and contingent. When I'm nice to a bunny rabbit I'm not being nice to bunny rabbit parasites. Amazing violence would be required to try to fit a form over everything all at once. If you try then you basically undermine the bunnies and everything else into components of a machine, replaceable components whose only important aspect is their existence. I assume you are sensitively aware of the ecological emergency we call the present—which has been happening in various forms for twelve thousand years. It is that there are logical limits on caring, a function of interdependence. Even the bodhisattva Avalokiteshvara couldn't save all sentient beings at once. This is why his head exploded into a thousand heads. That's what compassion (which is the beauty feeling) feels like. It is here that we encounter a deeper laughter than the laughter of absurdity. The laughter is about feeling a thing but being unable to grasp it or of knowing something but being unable to describe it. These pairs contradict, but they go together. This isn't just because of (human) perception or reason. There is a region of entities that support the loops of so-called subjects, their Harold and the Purple Crayon fantasies of being The Decider. Such entities are also in a loop. The style of a thing is always the slapstick impersonation of a thing. Toys are in an irreducible loop between themselves and themselves. Toys in this region are *silly*. Here we arrive at a truly comic level, the breadth of coexistence. Giddy laughter begins to break out, *because of The Sadness*.

Interdependence is the deep reason why at high resolution the language of rights breaks down for imagining how to proceed ethically and politically with regard to nonhumans. Extending rights to everything is absurd since rights language is normative: some beings can have rights to the extent that others do not. Rights language cancels itself out or leads to marginal cases that we humans are once again obliged to police. And, since rights language is based on property, it derives from one of the virulent lines of agrilogistic code. If everything has rights, then nothing can be property, for the same reason that rights language is normative. Furthermore, the one thing one can't possess in the ontological sense is oneself. One finds oneself "possessed," rather, by all kinds of (other) beings.

Universal rights are undermined by the notions of finitude and the hyperobject. The human as a species is not a universal but a very large finitude, absolutely concrete. It is better to think with Feuerbach that the human is not an abstract category but an actually existing being residing at a very large scale.

For all these reasons there are *already* no universal (human) rights, international declarations notwithstanding; a fact that is sadly borne out by the evidence. And although rights language is good enough to be getting on with if we want to save chimps (let's say) from being imprisoned, the fact that there is always one missing piece when we think interdependence, or act with a regard to it, means that we simply can't extend rights to all nonhumans all at once. So it's shocking to say that AIDS has just as much a right to exist as a human being—but luckily it's nonsense. Kindness cannot be perfectly automated.

The Longing. Inside that sadness liquid we reach a region of longing. Narcissism opens this possibility space. In The Sadness we encounter love for no reason—unconditional love. This is not different from longing, not a fullness denied by the supposed shallowness of longing. Need and desire definitely mean the same thing down here: "I need you." Why long for a polar bear or a forest or indeed a human? There is no good reason. Once you have enumerated good reasons to your satisfaction, the forest has burned down.[83] The polar bear has drowned. The movement down here toward the center of the chocolate is from compassion to passion, the possibility condition for compassion. Otherwise compassion is simply "idiot compassion," a condescending pity that maintains existing power relations: "Pity would be no more / If we did not make somebody poor."[84] We have anxiety because we care; it's a caring chocolate. We turn toward the chocolate because of longing, though it may look like guilt at first and descend through the subsequent guises, many layers of care. Longing is like depression that melted. The laughter of longing is a laughter of released energy no longer tied to a concept or an (objectified) object of any kind, free floating, amazed at its abundance.

The Longing is spiritual in the sense that it isn't cynical reason, insofar as it takes consciousness more seriously than seeing it as a reified instrument. Yet it is nonreligious in that it is committed to knowing reality or, better, those who are real. It is nontheistic, which isn't the same as atheist. Thinking down here requires that we relax the Law of the Excluded Middle. The boundary between sadness and longing is undecidable. Bittersweet, like good chocolate. And it means that we relax any inhibitions about being *spiritual,* that dreaded New Age term.

The Joy. Why longing? Because of joy. The basic toylike nature of things means that reality fundamentally is playful, dancing, raving, elemental. This is hard to accept in a depressing ecological age and more generally during the time of agrilogistics, a social form with a depressively narrow temporality diameter. So perhaps we should explore comedy and laughter a little before we proceed. The laughter of joy is full-on utter hilarity, accurately tracking ontological hilarity. Art begins to sound like dance music.

To locate the pathway toward The Joy, we will need to examine how things can become too serious. When you are funny it means that you allow the irreducible gap between what you are and who you think you are to manifest without tampering with it. When you are successfully funny, it means you allow people to see you being that, living that gap. You are radically accepting your finitude.[85] Depression is an autoimmune disorder of the intellect against its poor phenomenological host being, little you. The "tears of a clown" form of comedic depression is when the depression says, *I am not (just) a finite being*, a sentence that sounds suspiciously like the agrilogistic virus. The desire arises to be regarded as a "serious" actor whose irreducible gap is sealed. Like white blood cells, the intellect can't bear mortality and finitude. It wants you to live forever. It will eliminate every contradiction in its path to carry out this (absurd, impossible, destructive) mission. The "logical" conclusion to this path is the suicidal elimination of the host, like going into anaphylactic shock.

The agricultural logistics that now dominates Earth is this depression mind manifesting in global space. Objectively eliminating the finitude and anomalies that actually allow it to happen, the poor voles and weeds. The level of ecological awareness after guilt and shame has to do with depression, of being de-pressed by the overwhelming presence of processes and entities that one can't shake off. The idea that one *could* shake them off is the basis of the depression. The depression is in effect a symptom of agrilogistics, itself a depressive drive to eliminate contradiction, with its consequent absurd and violent demarcation of Nature and (human) culture. Depression in a box, Mesopotamian depression, obsessively reproduced, now global. The whole point is to fight one's way back from the brink (species-cidal and suicidal) toward the comedy. Toward accepting the irreducible rift between what a thing is and how it appears, allowing it to manifest.

The neurologist Adam Kaplin asserts, "The worst part of depression is that it narrows the field of vision into a very small tube so they can't see the options."[86] Maximum tube compression as far as my experiences of depression have been concerned has consisted of five minutes into the future and five minutes into the past. Humans find it hard to survive if their temporality is restricted to a diameter of ten minutes. Again, there is an ecological resonance here: agrilogistics compresses temporality to diameters that are dangerous to lifeforms, including humans, and how we inhabit Earth and coexist with other beings affects us too. "Each year, 34,000 people commit suicide, about twice as many deaths as caused by homicide—about one death per 15 minutes. By 2030, depression will outpace cancer, stroke, war and accidents as the world's leading cause of disability and death, according to the World Health Organization."[87] Thinking that you or they can snap out of it is addiction speak akin to what Gregory Bateson calls the "heroic" style of alcoholism: *I can master myself.*[88] The trouble is that this thought *is itself depression*.

Agrilogistics is Easy Think spacetime. A one-size-fits-all depression temporality, a sad rigid thin gray tube. We are living inside depression objectified in built space. It's in the way gigantic fields of rapeseed extend everywhere. It's in the way huge lonely front lawns extend a meaningless one-size-fits-all statement about individuality. It's in the way malls have gigantic parking lots, and housing lots have giant McMansions without so much as a garden. With its tiny temporality window, agrilogistic depression has turned the surface of Earth into a fatal place. Not only the land but also the oceans, which are the unconscious of the built space, the toilet where the chemicals go. As we have seen, there is a simple Freudian term for a fatal compulsion that keeps on retweeting: *death drive*.

Now to think the Joy, we simply invert these parameters. Instead of the fatal game of mastering oneself, ecognosis means realizing the irony of being caught in a loop and how that irony does not bestow escape velocity from the loop. Irony and sincerity intertwine. This irony is joy, and the joy is *erotic*. As Jeffrey Kripal puts it, gnosis is *thought having sex with itself*.[89] This is not a dance in the vacuum of an oukontic nothing. Eros is an attunement, and if there is attunement there is an *already-being*. A dance that knows itself: unlike the patriarchal "Woman," a *chora* (container) who cannot know herself as such, ecognosis is a *chora* who can.[90]

"Something" is "there": the elemental givenness of the arche-lithic. The arche-lithic isn't a space where relations between distinct separable beings are what makes them real. Relation "between a being"—relation between a being and itself—is the possibility condition for any other relating. The attunement itself is a not-me such that when thought is attuning to itself there is already a being, though I can't grasp it. The warm safety of The Sadness depends on the safety and care of The Longing, which in turn depends on the basic effervescence of The Joy, the welling up of an already-being. This attunement is itself ecological because Joy functions without me. There is no (mundane, objectified) reality outside the loop. The loop as such is a dance, a pattern in excess of what is patterned, pointing to infinity. This Joy is not despite the tree, the seagull, the lichen; it is the elixir of their finitude.

In a sense, *all toys are sex toys* to the extent that they enable links between beings and *between a being*. The threatening corniness of James Cameron's *Avatar* reaches a peak in the living devices that connect the Navi to the biospheric Internet, as when they plug their tails into the skulls of flying lizards.[91] Is it anything other than needless to point out the eroticism? The erotic wiring together of beings suggests the wiring between a being: the ultimate gnosis in *Avatar* would be to plug the tail into oneself . . . In The Joy there is an excess of links between a being over links between different beings. Is it too ungrammatical to say *between the same being*? Between the being that is oneself, even between thinking and itself. Although cloning is chronologically prior to sex, perhaps sex is logically prior to cloning. We consider here certainly not a heteronormative sex, but sex for its own sake whose prototype is denigrated as narcissistic. Buddhist Tantra provides a template: ultimate reality is seen as emptiness (the radical inaccessibility of things) in sexual union with appearance (their shimmering givenness), different but the same.

Esoteric regions in Axial Age religions, often persecuted by the ortho-dox (the straighteners), seem to evoke something of the arche-lithic, the timeless time of coexistence and ontological confusion and profusion. By contrast, the joyless agrilogistic reproduction of existence for its own sake is *heterosexuality as cloning*. Only consider the prevalence of agricultural images for the worst kinds of sex, replete with plowings and wild oats. Recall the clonelike endlessness of suburban sprawl, its relentlessly conformist agrilogistic format. Remember that agrilogistics is a loop and a "for its own sake" that denies that aestheticism—a denial whose function is to reproduce one toy, one political form over and over again, a megatoy that puts an end to play.

It would be a grave mistake (the mistake of a terrible seriousness) to try and get rid of the "for its own sake," which is nothing other than the intrinsic loop form of being, let alone of ecological being, let alone the repressed (hence dangerous) form of agrilogistics. This loop form is why we suspect the aesthetic. The cry of "We have to *do* something!" defining itself militantly against a reified notion of not-doing, means "At all costs we must interrupt the loop." One needs to delve *further into* the loop form, into the scary regions of consumerism, aestheticism, narcissism. As Adorno pointed out, the exit route looks like a regression.[92] The arche-lithic appears as decadence and so is ignored. Let's not interrupt the loop. Let's interrupt the violence that tries to straighten the loop. In this light, it's not just that the heteronormativity of some environmentalist speech needs to be challenged out of nowhere, because it is not congruent with present understanding, ethics, and politics. It is that this *heteronormative environmentalism is agrilogistic*, and as such it masks the arche-lithic basis of ecological coexistence.

The Joy is logically prior to life, deep inside life, the quivering between two deaths. Deep in the interior of life there are dancing puppets. In the same way that viruses are logically prior to bacteria, thoughts are logically prior to minds, hallucinations are logically prior to thoughts, flowers are logically prior to plants, patterns are logically prior to evolution.[93] In Dick's *A Scanner Darkly* the little blue flower from which Substance D is the extract is prior to its "active ingredient." A little blue flower, so strangely not part of a recognizable agrilogistic plan, which was to shrink flowers and maximize juicy, substance-producing kernels. To eliminate the disturbing playfulness of patterns. The fact that a molecule can make a pattern is a possibility condition for its being able to self-replicate, which then initiates the organism–environment dance that is evolution. Something radically nonutilitarian, outside "Life," bankrolls evolution's utilitarian appearance with its play, empathy, and mutual aid. Life is already a kind of death drive, a putting-something-to-work, taking advantage of a molecule's ability to generate patterns for patterns' sake. Something radically nonutilitarian is a possibility condition for the "work" of evolution, culture and agriculture, steam engines and the adult world. Children's play does not adhere to a fake, narrow-bandwidth version of itself, which pretends to be "reality" that puts away childish things.

In fully realized ecognosis the chocolate has been turned inside out. A tiny crystal of guilt sugar is contained within a little ball of shame enveloped in a congealed sphere of melancholia swimming in a galaxy of sadness contained within a plasma field of joy. This plasma field is a Ganzfeld effect of affect, as in a blizzard or a light installation by James Turrell, where one's sense of distance evaporates. When

I stand in a Turrell exhibit, the environment at its purest seems to absorb me from all sides, without objects—or, rather, the environment itself has become one gigantic object, not simply a background or blank slate or empty stage set or envelope.[94] Not even what in some phenomenological research is called *the surround*.[95] What precisely is it surrounding? That is the whole question. I find myself thrown out of my habitual sense of where I stop and start just as much as the curving walls and soft yet luminous colors melt the difference between *over here* and *over there*. A double invagination: first the reified art object is opened, its givenness allowed to permeate everywhere. Then this opening is itself opened, and we find ourselves, weirdly, on the inside of an entity, an uncanny entity that we cannot grasp, yet one that is palpable, luminous, exactly this shade of pink.[96]

Abjection has been transfigured into what Irigaray calls *nearness*, a pure givenness in which something is so near that one cannot *have* it—a fact that obviously also applies to one's "self."[97]

The Joy is not abstract or uniform, but so intimate you can't see it, and you can't tell whether it's inside or outside: the "cellular" experience of bonds tightening between beings.[98] It's not an emotion that I'm having. I'm in a passion. A passion is not in me.[99] The Ganzfeld effect of The Joy is haptic, elemental: so close that you lose track of something to be seen. Here thought itself is a way of getting high: human attunement to thinking has been intoxicated into recognizing its nonhuman status. Not simply thinking ecologically (the ecological thought), but rather thought as susceptibility, thinking as such as ecology. The structure of thought as nonhuman. Ecognosis.

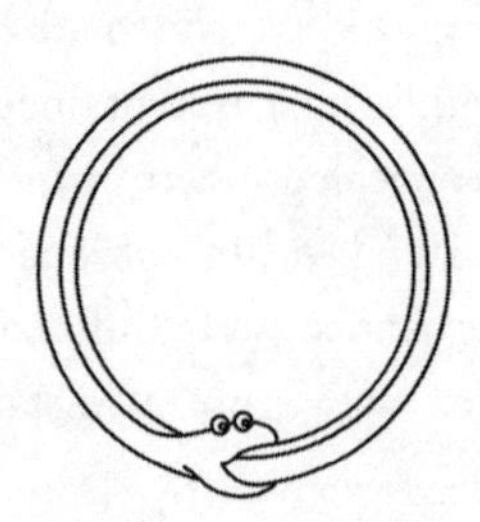

ENDING BEFORE THE BEGINNING

The ecological era we find ourselves in—whether we like it or not and whether we recognize it or not—makes necessary a searching revaluation of philosophy, politics, and art. The very idea of being "in" an era is in question. We are "in" the Anthropocene, but that era is also "in" a moment of far longer duration.

What is the present? How can it be thought? What is presence? Ecological awareness forces us to think and feel at multiple scales, scales that disorient normative concepts such as "present," "life," "human," "nature," "thing," "thought," and "logic." *Dark Ecology* shall argue that there are layers of attunement to ecological reality more accurate than what is habitual in the media, in the academy, and in society at large.

These attunement structures are necessarily *weird*, a precise term that we shall explore in depth. Weirdness involves the hermeneutical knowingness belonging to the practices that the humanities maintain. The attunement, which I call *ecognosis*, implies a practical yet highly nonstandard vision of what ecological politics could be. In part, ecognosis involves realizing that nonhumans are installed at profound levels of the human—not just biologically and socially but in the very structure of thought and logic. Coexisting with these nonhumans is ecological thought, art, ethics, and politics.

Dark Ecology traces the ecological crisis to a logistical "program" that has been running unquestioned since the Neolithic. *Dark Ecology* argues that ecological reality requires an awareness that at first has the characteristics of tragic melancholy and negativity, concerning inextricable coexistence with a host of entities that surround and penetrate us, but which evolves paradoxically into an anarchic, comedic sense of coexistence. In the First Thread of this book, I argue that ecological awareness takes the form of a loop. In this loop we become aware of ourselves as a species—a task far more difficult than it superficially appears. We also grow familiar with a logistics of human social, psychic, and philosophical space, a twelve-thousand-year set of procedures that resulted in the very global warming it was designed to fend off. The Second Thread shows that the logistics represses a paradoxical realm of human-nonhuman relations. The realm contains tricksterlike beings that have a loop form, which is why ecological phenomena and awareness have a loop form. The growing familiarity with this state of affairs is a manifestation of dark ecology. Dark ecology begins in darkness as depression. It traverses darkness as ontological mystery. It ends as dark sweetness. The Third Thread maps these stages, while outlining the ethics and politics that emerge from dark ecology.

The Arctic Russian town of Nikel looks horrifying at first, like something out of Tarkovsky's *Stalker,* only on bad acid. A forest devastated by an iron-smelting factory. Soviet buildings stark and bleak. Mounds of garbage sitting on hills of slag. A final tree, last of the pines destroyed by the sulfur dioxide. We were Sonic Acts, a small group of musicians, artists, and writers. We had traveled there in later 2014 to start a four-year art and research project called *Dark Ecology.*

Then Nikel becomes rather sad and melancholic. A collection of broken things. Past things. Garages repurposed as homes. Broken metal structures in which people are living. Holding on to things for no reason. Sometimes the smelting plant is closed because the price of iron plummets. Then all lose their jobs. Then it restarts. The Norwegians pay the plant to direct its smoke elsewhere than across the very border, ever so close. That means the factory directs the smoke over Russians. Peeling paint, telling stories of decisions and indecisions and nondecisions.

And then for some strange reason it becomes warm. There is a Palace of the Future, full of wonderful kitschy communist art, Terry Gilliam sculpturelike lampshades, hauntingly luminous pale blues, pinks, and yellows, the building grooving as hard as a Tibetan stupa. And on the outskirts the reality of death is so explicit. It's a charnel ground almost identical to the one on Mount Kailash, another very friendly place where offerings (or are they huge piles of garbage?) litter the space at the top and nuns meditate in a land strewn with bits of corpse like an emergency room. People are dying, or are they going to live, or are they already dead? There is a lot of blood, severing and severed limbs. A lot of care.

It's even a little bit funny. A drag queen poses for a photographer outside a metallic building. Some kind of joy is here. The demons and ghosts aren't demons or ghosts. They are faeries and sprites. The arche-lithic.

Dark ecology thinks the truth of death, a massive cognitive relief that if integrated into social form would embody nonviolence. It makes you wonder, maybe we should store plutonium neither deep underground with militarized warnings nor in knives and forks without any warning whatsoever (this was actually suggested in the late 1990s). Let's get small pieces of plutonium, store them in a way that we can monitor them, and encase them in a substance that will not leak radiation, aboveground, so you can maintain the structure and so that you can take responsibility for it. You, the human, made the plutonium, or you the human can understand what it is—therefore you are responsible. Let's put these structures in the middle of every town square in the land. One day there will be pilgrimages to them and circumambulations. A whole spirituality of care will arise around them. Horror and depression will give way to sadness and joy. We bristle plutoniumly. Or we feel suicidal plutoniumly. Or we cry plutoniumly. Or we even dance plutoniumly. The arche-lithic. There is always already a relationship to a nonhuman.

In anticipation of this future, let's make metal personnel covers and plant them in the town square. On the covers will be stamped the following sentence: FUTURE PEOPLE WILL MAINTAIN PLUTONIUM HERE. Or let's team up with some physicists and get hold of a plutonium battery. Let's encase it in a safe storage chamber. Then let's put it in MoMA. That might be a good start.

Jae Rhim Lee's Mushroom Burial Suit is infused with familiar mushroom mycelia. With hairs and skin flakes and tweezers and petri dishes, you can train these mushrooms to recognize and enjoy eating your flesh. Then, when you die, you can be placed in your suit in one of her flat-pack slotted cardboard coffins, and the mushrooms digest you in two days. It's even better than being left for the vultures, because the mushrooms metabolize the mercury. Or you could act like an Egyptian seed and be cryogenically suspended for several hundred years—when the future people open the chamber, they will be so pleased to see you. How did they manage without you? They will regard the fact that the chamber ate even more energy than regular living you as a minor inconvenience. Decomponaut or cryonaut: it depends on whether you are OK with widening your view, taking your eyes out of the telescopic sight of Life, putting down the agrilogistic tube, and resting in the charnel ground.[1]

Let's pour the oil of death on the troubled water of agrilogistics. Let's disco.

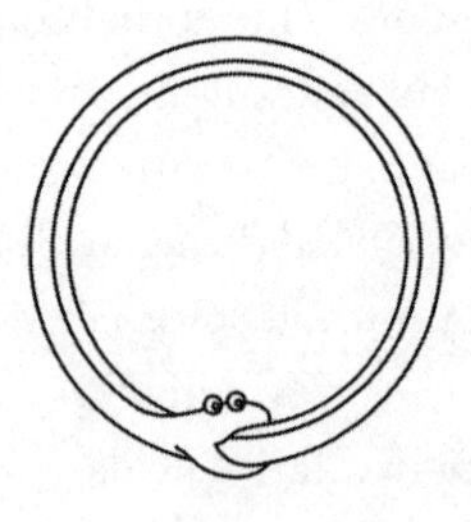

NOTES

THE FIRST THREAD

1. Thomas Hardy, *Tess of the D'Urbverilles,* ed. David Skilton, intro. A. Alvarez (Harmondsworth: Penguin, 1984), 136–39.
2. Timothy Morton, *Hyperobjects: Philosophy and Ecology After the End of the World* (Minneapolis: University of Minnesota Press, 2013).
3. Tom Engelhardt, "The 95% Doctrine: Climate Change as a Weapon of Mass Destruction," openDemocracy, May 27, 2014, http://www.opendemocracy.net/print/83094 (accessed June 11, 2014).
4. In 2013 Paul Kingsnorth published an essay called "Dark Ecology: Searching for Truth in a Post-Green World" in *Orion* magazine (January-February 2013). *Dark ecology* is a term I had coined in 2004 and explored in *Ecology Without Nature: Rethinking Environmental Aesthetics* (Cambridge: Harvard University Press, 2007).
5. *Oxford English Dictionary, weird,* adj. http://www.oed.com (accessed April 9, 2014).
6. S. N. Hagen, "On Nornir 'Fates,'" *Modern Language Notes* 39, no. 8 (December 1924): 466–69.
7. *Oxford English Dictionary*, "weird," n. 1.a., 1.b., 2.a, http://www.oed.com (accessed October 4, 2012).
8. *Oxford English Dictionary*, "worth," v.1, http://www.oed.com (accessed October 4, 2012).
9. *Oxford English Dictionary*, "weird," adj. 1, 2.a., 3, http://www.oed.com (accessed December 11, 2013).
10. *Oxford English Dictionary, faerie, fay* n.2, http://www.oed.com (accessed September 25, 2014).

11. C. S. Holling and Gary K. Meffe, "Command and Control and the Pathology of Natural Resource Management," *Conservation Biology* 10, no. 2 (April 1996): 328–37.
12. Michael Wines, "Mystery Malady Kills More Bees, Heightening Worry on Farms," *New York Times*, March 28, 2013, http://www.nytimes.com/2013/03/29/science/earth/soaring-bee-deaths-in-2012-sound-alarm-on-malady.html?pagewanted=all&_r=0 (accessed March 31, 2013). Brad Plumer, "We've Covered the World in Pesticides: Is That a Problem?" *Washington Post*, August 18, 2013, http://www.washingtonpost.com/blogs/wonkblog/wp/2013/08/18/the-world-uses-billions-of-pounds-of-pesticides-each-year-is-that-a-problem/ (accessed August 25, 2013).
13. Suzanne Goldenberg, "Americans Care Deeply About 'Global Warming'—But Not 'Climate Change,'" *Guardian*, May 27, 2014, http://www.theguardian.com/environment/2014/may/27/americans-climate-change-global-warming-yale-report/print (accessed June 2, 2014).
14. Lewis Carroll, *Alice Through the Looking Glass* in *The Annotated Alice: The Definitive Edition*, ed. Martin Gardner (New York: Norton, 2000), 157.
15. Paul Crutzen and Eugene Stoermer, "The Anthropocene," *Global Change Newsletter* 41, no. 1 (2000): 17–18.
16. Elizabeth Kolbert, *The Sixth Extinction: An Unnatural History* (New York: Holt, 2014).
17. Shu-zhong Shen et al., "Calibrating the End-Permian Mass Extinction," *Science Express*, November 17, 2011, www.sciencexpress.org / 17 November 2011 / Page 1 / 10.1126/science.1213454 (accessed November 20, 2011).
18. Dipesh Chakrabarty, "The Climate of History: Four Theses," *Critical Inquiry* 35 (Winter 2009): 197–222 (206–7).
19. The view I am calling out of date is expounded in Peter Sloterdijk, *In the World Interior of Capital* (Cambridge: Polity, 2014).
20. An exemplary passage is found in Sloterdijk himself, ibid., 251–54.
21. I put a hyphen between *post* and *modern* in the manner of Iain Thomson. See *Heidegger, Art, and Postmodernity* (Cambridge: Cambridge University Press, 2011); "Heidegger's Aesthetics," in Edward N. Zalta, ed., *The Stanford Encyclopedia of Philosophy* (Summer 2011 edition), http://plato.stanford.edu/entries/heidegger-aesthetics/, November 30, 2014.
22. The highly necessary fantasy component of this imaginary is found in premodern Europe, complicating the view expounded most recently by Sloterdijk (and by a host of others, for instance Deleuze) that the drive to homogeneity and speed is a symptom of the modern. See Timothy Morton,

The Poetics of Spice: Romantic Consumerism and the Exotic (Cambridge: Cambridge University Press, 2000).

23. See Morton, *Hyperobjects.*
24. Michel Foucault, *The Order of Things: An Archaeology of the Human Sciences* (New York: Random House, 1994), 387.
25. I refer to the action performed by the government of the Maldives in 2009.
26. Quentin Meillassoux, *After Finitude: An Essay on the Necessity of Contingency*, trans. Ray Brassier (New York: Continuum, 2009), 5.
27. IPCC (Intergovernmental Panel on Climate Change), *Climate Change 2014: Summary for Policymakers*, http://www.ipcc.ch/pdf/assessment-report/ar5/syr/SYR_AR5_SPM.pdf (accessed November 23, 2014).
28. Chakrabarty, "The Climate of History."
29. Genetic affinities between humans and Neanderthals are becoming more obvious every day. Carl Zimmer, "Neanderthals Leave Their Mark on Us," *New York Times*, January 29, 2014, http://www.nytimes.com/2014/01/30/science/neanderthals-leave-their-mark-on-us.html?hpw&rref=science&_r=0 (accessed January 31, 2014).
30. Timothy Morton, *Realist Magic: Objects, Ontology, Causality* (Ann Arbor: Open Humanities, 2013).
31. Meillassoux and Ray Brassier hold this position.
32. Hence the charm of Neil Shubin's documentary *Your Inner Fish* (PBS, 2014).
33. Charles Darwin, *The Origin of Species*, ed. Gillian Beer (Oxford: Oxford University Press, 1996), 160.
34. Jacques Derrida, "Hostipitality," trans. Barry Stocker with Forbes Matlock, *Angelaki* 5, no. 3 (December 2000): 3–18; Timothy Morton, *The Ecological Thought* (Cambridge: Harvard University Press, 2010), 14–15, 17–19, 38–50.
35. Sesame Street, "We Are All Earthlings," *Sesame Street Platinum All-Time Favorites* (Sony, 1995); USA for Africa, "We Are the World" (Columbia, 1985).
36. See for instance Kim Stanley Robinson, *Red Mars* (New York: Random House, 1993), *Green Mars* (New York: Random House, 1995), *Blue Mars* (New York: Random House, 1997).
37. Carroll, *Alice Through the Looking Glass,* 187.
38. James Lovelock, *The Revenge of Gaia: Earth's Climate Crisis and the Fate of Humanity* (New York: Basic Books, 2006), 6–7.
39. Douglas Kahn, *Earth Sound Earth Signal: Energies and Earth Magnitude in the Arts* (Berkeley: University of California Press, 2013).

40. Bruno Latour, "Anthropology at the Time of the Anthropocene: A Personal View of What Is to be Studied," http://www.bruno-latour.fr/sites/default/files/139-AAA-Washington.pdf (accessed May 12, 2015).
41. This idea is occurring to a number of people simultaneously. See for instance Charles C. Mann, "State of the Species: Does Success Spell Doom for Homo Sapiens?" *Orion* (November-December 2012), http://www.orionmagazine.org/index.php/articles/article/7146 (accessed November 14, 2012).
42. I use the term *ontic* as Martin Heidegger uses it in *Being and Time*, trans. Joan Stambaugh (Albany: State University of New York Press, 2010), 11.
43. Naomi Klein, "Climate Change Is the Fight of Our Lives—Yet We Can Hardly Bear to Look at It," *Guardian,* April 23, 2014, http://www.theguardian.com/commentisfree/2014/apr/23/climate-change-fight-of-our-lives-naomi-klein/print (accessed April 26, 2014).
44. I'm grateful to my talented Ph.D. student Toby Bates for pointing this out.
45. Charles Long, *Catalin*, The Contemporary, Austin, March 2014.
46. John Bellamy Foster, *Marx's Ecology: Materialism and Nature* (New York: Monthly Review Press, 2000); "Marx's Ecological Value Analysis," *Monthly Review* (September 2000): 39–47.
47. This narrowness affects definitions of *materialism,* for instance in Alexander Galloway, "The Poverty of Philosophy: Realism and Post-Fordism," *Critical Inquiry* 39, no. 2 (Winter 2013): 347–66. Materialism in this sense actually means human economic metabolism.
48. Sloterdijk, *Interior*, 223–32 (224).
49. Ibid., 230.
50. *Oxford English Dictionary*, "cattle," n.I.1, 2.a., 3, 4, http://www.oed.com (accessed December 6, 2014).
51. Karl Marx, *Capital*, trans. Ben Fowkes (Harmondsworth: Penguin, 1990), 1:283–84 (chapter 7).
52. T. Sasaki and S. C. Pratt, "Ants Learn to Rely on More Informative Attributes During Decision-Making," *Biology Letters* 9, no. 6 (2013): DOI: 10.1098/rsbl.2013.0667.
53. Jessica Morrison, "Bees Build Mental Maps to Get Home," *Nature News,* June 2, 2014, http://www.nature.com/news/bees-build-mental-maps-to-get-home-1.15333 (accessed June 11, 2014).
54. David McNamee, "Rats 'Experience Regret, Too,'" *Medical News Today,* June 9, 2014, http://www.medicalnewstoday.com/articles/277959 (accessed July 15, 2015).

55. Stuart H. Hurlbert, "Pseudoreplication and the Design of Ecological Experiments," *Ecological Monographs* 54, no. 2 (1984): 187–211.
56. Morton, *Realist Magic.*
57. Michael Suk-Young Chwe, "Scientific Pride and Prejudice," *New York Times*, January 31, 2014.
58. Martin Heidegger, *What Is a Thing?* trans. W. B. Barton and Vera Deutsch, analysis by Eugene T. Gendlin (Chicago: Henry Regnery, 1967).
59. Damian Carrington, "Earth Has Lost Half of Its Wildlife in the Past 40 Years, Says WWF," *Guardian*, September 29, 2014, http://www.theguardian.com/environment/2014/sep/29/earth-lost-50-wildlife-in-40-years-wwf/print (accessed September 30, 2014).
60. Addy Pross, "Toward a General Theory of Evolution: Extending Darwinian Theory to Inanimate Matter," *Journal of Systems Chemistry* 2, no. 1 (2011): 1–14. For a study of the problems of replicator theories, see Robert Shapiro, "A Replicator Was Not Involved in the Origin of Life," *Life* 49 (2000): 173–76.
61. Immanuel Kant, *Critique of Pure Reason,* trans. Norman Kemp Smith (New York: St. Martin's, 1965), 84–85, A45–A47/B63–64.
62. See for example the "you're either with us or against us" rhetoric of Galloway, "The Poverty of Philosophy."
63. Eric Posner and David Weisbach, "Public Policy Over Massive Time Scales," The History and Politics of the Anthropocene, University of Chicago, May 17–18, 2013.
64. Sean Hernandez, "The Darkest of Greens: Measuring the Incidence and Character of Eco-Depression in Undergraduates," paper given at Approaching the Anthropocene: Perspectives from the Humanities and Fine Arts, May 7–8, 2015.
65. Gilbert N. Plass, "The Carbon Dioxide Theory of Climate Change," *Tellus* 8, no. 2 (1956): 140–54 (143–44, 149–50).
66. Gifford H. Miller et al., "Unprecedented Recent Summer Warmth in Arctic Canada," *Geophysical Research Letters* 40, no. 21 (November 16, 2013): 5745–51.
67. Sigmund Freud, *The Uncanny*, trans. David McClintock and Hugh Haughton (London: Penguin, 2003).
68. For an exhaustive account of the ethical dilemmas that present themselves during the time of global warming, see Stephen M. Gardiner, *A Perfect Moral Storm: The Ethical Tragedy of Climate Change* (Oxford: Oxford University Press, 2011).

69. Kelly Levin, Benjamin Cashore, Graeme Auld, and Steven Bernstein, "Playing It Forward: Path Dependency, Progressive Incrementalism, and the 'Super Wicked' Problem of Global Climate Change," *IOP Conference Series: Earth and Environmental Science*, 6, session 50, DOI: *10.1088/1755–1307/6/50/502002*,http://iopscience.iop.org/1755–1315/6/50/502002/ (accessed May 1, 2015).
70. *The Ramayana of Valmiki*, trans. M. L. Sen (New Delhi: Munshiram Manoharial, 1976).
71. Genesis 3:17–19 (New Living Translation).
72. See the marvelously exhaustive collection of data in Mark Nathan Cohen and George J. Armelagos, *Paleopathology at the Origins of Agriculture* (Gainsville: University Press of Florida, 2013).
73. Joseph Russell Smith, *Tree Crops: A Permanent Agriculture* (New York: Harcourt, Brace, 1929), 4.
74. Jes Benstock and Luke Losey, video for Orbital, "The Box" (Technobabble, 1996).
75. Blake Hurst, "Big Farms Are About to Get Bigger," *American*, December 11, 2013, http://www.american.com/archive/2013/december/big-farms-are-about-to-get-bigger/article_print (accessed December 14, 2013).
76. Andrew Curry, "The Milk Revolution," *Nature News and Comment*, July 31, 2013, http://www.nature.com/news/archaeology-the-milk-revolution-1.13471 (accessed October 9, 2013).
77. There are far too many texts to mention, but two reasonably recent ones that have stood out for me have been Geoffrey Hartman, *The Fateful Question of Culture* (New York: Columbia University Press, 1997) and Terry Eagleton, *The Idea of Culture* (Oxford: Blackwell, 2000).
78. In New Guinea, native pigs can't plough—so agrilogistics was stymied there again.
79. Jan Zalasiewicz, "The Geological Basis for the Anthropocene," The History and Politics of the Anthropocene, University of Chicago, May 17–18, 2013.
80. Jared Diamond, "The Worst Mistake in the History of the Human Race," *Discover Magazine* (May 1987): 64–66. Derek Parfit, *Reasons and Persons* (New York: Oxford University Press, 1984). He offers a slightly revised discussion in "Overpopulation and the Quality of Life," in *Applied Ethics,* ed. Peter Singer (New York: Oxford University Press, 1986). Giorgio Agamben, *Homo Sacer: Sovereign Power and Bare Life* (Stanford: Stanford University Press, 1998).

81. On the patriarchy aspect insofar as it affects philosophy as such, Luce Irigaray is succinct: woman has been taken "*quoad matrem* . . . in the entire philosophic tradition. It is even one of the conditions of its possibility. One of the necessities, also, of its foundation: it is from (re)productive earth-mother-nature that the production of the logos will attempt to take away its power, by pointing to the power of the beginning(s) in the monopoly of the origin." *This Sex Which Is Not One*, trans. Catherine Porter and Carolyn Burke (Ithaca: Cornell University Press, 1985), 102.
82. *Shelley's Poetry and Prose*, ed. Donald H. Reiman and Neil Fraistat (New York and London: Norton, 2002).
83. See, for instance, Pedro Barbosa, ed., *Conservation Biological Control* (San Diego: Harcourt Brace, 1998).
84. Tom Stoppard, *Darkside: A Play for Radio Incorporating the Dark Side of the Moon* (Parlophone, 2013).
85. Rebecca J. Rosen, "How Humans Invented Cats," *Atlantic*, December 16, 2013, http://www.theatlantic.com/technology/archive/2013/12/how-humans-created-cats/282391/ (accessed November 21, 2014); Gerry Everding, "Cat Domestication Traced to Chinese Farmers 5,300 Years Ago," *Washington University St. Louis Newsroom,* December 16, 2013, https://news.wustl.edu/news/Pages/26273.aspx (accessed November 21, 2014; Carlos A. Driscoll, "The Taming of the Cat," *Scientific American* 300, no. 6 (June 2009): 68–75; Yaowu Hu et al., "Earliest Evidence for Commensal Processes of Cat Domestication," *PNAS* 111, no. 1 (January 7, 2014): 116–20.
86. See, for instance, Donna Haraway, *When Species Meet* (Minneapolis: University of Minnesota Press, 2007).
87. For arguments in support of this hypothesis, see Terry O'Connor, *Animals as Neighbors: The Past and Present of Commensal Animals* (East Lansing: Michigan State University Press, 2013).
88. Victoria Scaven and Nicole Rafferty, "Physiological Effects of Climate Warming on Flowering Plants and Insect Pollinators and Potential Consequences for Their Interactions," *Current Zoology* 59, no. 3 (2013): 418–26. Christoph Sandrock et al., "Sublethal Neonicotinoid Insecticide Exposure Reduces Solitary Bee Reproductive Success," *Agricultural and Forest Entomology* 16 (2014): 119–28. "U.S. Beekeepers Lost 40 Percent of Bees in 2014–15, *Science* Daily, May 21, 2015, http://www.sciencedaily.com/releases/2015/05/150513093605.htm (accessed July 21, 2015). Michael Kuhlmann, Danni Guo, Ruan Veldtman, and John

Donaldson, "Consequences of Warming Up a Hotspot: Species Range Shift Within a Centre of Bee Diversity," *Diversity and Distributions* 18 (2012): 885–97.

89. Philip Roth, *Nemesis* (New York: Vintage, 2011).
90. Joseph Nesme et al., "Large-Scale Metagenomic-Based Study of Antibiotic Resisatnce in the Environment," *Current Biology* 24, nos. 1–5 (May 2014), http://dx.doi.org/10.1016/j.cub.2014.03.036 (accessed July 21, 2015), 1, 2–3.
91. Jacques Derrida, "Violence and Metaphysics," in *Writing and Difference*, trans. Alan Bass (London: Routledge, 2001), 162–66.
92. Stoppard, *Darkside*.
93. Richard Manning, "The Oil We Eat," *Harper's Magazine*, February 4, 2004, http://www.wesjones.com/oilweeat.htm (accessed February 17, 2014). See Richard Manning, *Against the Grain: How Agriculture Has Hijacked Civilization* (New York: North Point, 2005).
94. Gardiner, *Perfect Moral Storm*, 213–45.
95. Diamond, "The Worst Mistake," 64–66.
96. Erle Ellis, "Overpopulation Is Not the Problem," *New York Times*, September 13, 2013, http://www.nytimes.com/2013/09/14/opinion/overpopulation-is-not-the-problem.html?smid=tw-share&_r=0&pagewanted=print.
97. Agamben, *Homo Sacer*.
98. Parfit, *Reasons and Persons*, 433–41.
99. Robert M. May, "Necessity and Chance: Deterministic Chaos in Ecology and Evolution," *Bulletin of the American Mathematical Society* 32, no. 5 (July 1995): 291–308 (306).
100. Mary Daly, *Gyn/Ecology: The Metaethics of Radical Feminism* (Boston: Beacon, 1990), 40–46.
101. Philip K. Dick, *A Scanner Darkly* (New York: Houghton Mifflin Harcourt, 2011), 264.
102. Ibid., 274.
103. Ibid., 287.
104. Ibid., 97–98. See Theodor Adorno, "Progress," *Philosophical Forum* 15, nos. 1–2 (Fall-Winter 1983–1984): 55–70 (61–63).
105. Dick, *A Scanner Darkly*, 283.
106. Ibid., 276.
107. René Descartes, *Meditations and Other Metaphysical Writings*, trans. and intro. Desmond M. Clarke (London: Penguin, 2000 [1998]).

108. Eric Michael Johnson, "Fire Over Ahwahnee: John Muir and the Decline of Yosemite," *Scientific American*, August 13, 2014, http://blogs.scientificamerican.com/primate-diaries/2014/08/13/fire-over-ahwahnee-john-muir-and-the-decline-of-yosemit/ (accessed November 26, 2014).
109. Morton, *Ecology Without Nature*.
110. On the going together of machination (logistics) and lived experience, see Martin Heidegger, *Contributions to Philosophy (From Enowning)*, trans. Parvis Emad and Kenneth Maly (Bloomington: Indiana University Press, 1999), 95–96.
111. Dick, *A Scanner Darkly*, 275.
112. Ibid., 97–98.
113. Rodolphe Gasché, "Reading Chiasms: An Introduction," in Andrzej Warminski, *Readings In Interpretation: Hölderlin, Hegel, Heidegger* (Minneapolis: University of Minnesota Press, 1987), ix–xxvi (xx). In *A Scanner Darkly* much of the proof of Arctor's paranoia relies on evidence that the brain hemispheres can operate separately when severed. One way this is achieved in the lab is through the cutting of the optic chiasm (Dick, *A Scanner Darkly*, 120).
114. It is well accepted that concentrations of O_{18}, an oxygen isotope, track climate stability. O_{18} concentrations were remarkably stable from the start of agrilogistics until the start of the Anthropocene.
115. Jan Zalasiewicz, presentation at History and Politics of the Anthropocene, University of Chicago, May 2013.
116. I am grateful to Jan Zalasiewicz for discussing this with me.
117. This is a self-imposed stumbling block for the otherwise excellent and audacious Julian Jaynes. See Julian Jaynes, *The Origin of Consciousness in the Breakdown of the Bicameral Mind* (New York: Houghton Mifflin, 1976).

THE SECOND THREAD

1. Isaiah (2:4), Joel (3:10). See for instance the Heidrick Agricultural History Center, Gibson House Museum, 1962 Hays Lane, Woodland, CA 95616, http://www.aghistory.org (accessed December 8, 2014).
2. Jean-Joseph Goux, *Oedipus, Philosopher*, trans. Catherine Porter (Stanford: Stanford University Press, 1993), 18–24.
3. William Elliott, "Losing Alaska to the Name Itself: Elegy and Futurity in a Changing North," Ph.D. dissertation, University of California, Davis, 2014.

4. Schopenhauer links jokes to this Kantian gap. Arthur Schopenhauer, *The World as Will and Representation,* trans. E. F. J. Payne, 2 vols. (New York: Dover, 1969), 1:76–77.
5. In this respect I differ from Goux, who is determined to read the role of the Sphinx (and her subsequent forgetting) as a "castration" of (male) human potential to transcend its physical conditions. As fascinating as this is—Goux is in effect noting that the transcendence of progress is only an illusion and we are still caught in an incestuous deadlock with our physicality—it is as if Goux thinks that if Oedipus had been more traditional, more "manly," he would have done better. Agrilogistics 1.0 would be better than agrilogistics 2.0. Evidently I reject both versions.
6. Anne Carson and Bianca Stone, *Antigonick* (New York: New Directions, 2012), 20.
7. Martin Heidegger, *Introduction to Metaphysics*, trans. Gregory Fried and Richard Polt (New Haven: Yale University Press, 2000), 112–26.
8. See also Hélène Cixous, *The Laugh of the Medusa*, trans. Keith Cohen and Paula Cohen, *Signs* 1, no. 4 (Summer 1976): 875–93 (882).
9. Carla Lonzi, *Let's Spit on Hegel*, trans. Veronica Newman (first published in Italian by Rivolta Femminile, 1970), 13, available at secunda.tumblr.com (accessed March 4, 2013). Also in Paola Bono and Sandra Kemp, eds., *Italian Feminist Thought: A Reader* (Oxford: Blackwell, 1991), 40–59.
10. Françoise d' Eaubonne, *La féminisme ou la mort* (Paris: P. Horay, 1974).
11. Bruno Latour, *We Have Never Been Modern* (Cambridge: Harvard University Press, 1993), 30, 151. The most invaluable study of the variegated field of ecofeminism is Stacy Alaimo, *Undomesticated Ground: Recasting Nature as Feminist Space* (Ithaca: Cornell University Press, 2000).
12. See, for instance, John Zerzan, "The Catastrophe of Postmodernism," *Future Primitive Revisited* (Port Townsend, WA: Feral House, 2012), 64–90. The first demon named is the loop of "Consumer narcissism" (64). In contrast, Neanderthal mind was fully present to itself and to its environment in a pure, nondeviant circularity, compared to which even the pre-Neolithic divisions of labor and cave paintings seem like original sin: "Running on Emptiness: The Failure of Symbolic Thought," *Running on Emptiness: The Pathology of Civilization* (Los Angeles: Feral House, 2002), 1–16 (2–3).
13. Julian Jaynes, *The Origin of Consciousness in the Breakdown of the Bicameral Mind* (Boston: Houghton Mifflin, 1990).
14. Predicibly, Harari's argument depends on a metaphysics that separates the objective from the subjective. Despite the promising Jared Diamond–style

and object-oriented argument that staple crops domesticated humans rather than the other way around, the emphasis on the power of (human) cognition and the difference between cognizing and everything else propels Harari toward his transhumanist conclusion. The solution to the violent transcendence of the nonhuman is—more transcendence. Yuval Noah Harari, *Sapiens: A Brief History of Humankind* (New York: HarperCollins, 2015).

15. Oliver Sacks, *Hallucinations* (New York: Vintage, 2013), ix–xiv.
16. Sugarhill Gang, "Rapper's Delight" (Sugar Hill Records, 1979). On the phatic, see Roman Jakobson, "Closing Statement: Linguistics and Poetics," in Thomas A. Sebeok, ed., *Style in Language* (Cambridge: MIT Press, 1960), 350–77.
17. Martin Heidegger, *Being and Time*, trans. Joan Stambaugh (Albany: State University of New York Press, 1996), division 1, chapter 6 (169–211).
18. *Finding Nemo*, dir. by Andrew Stanton and Lee Unkrich (Burbank: Walt Disney and Pixar Animation, 2003).
19. Clifford N. Matthews, "The HCN World," *Cellular Origin, Life in Extreme Habitats and Astrobiology* 6 (2005): 121–35. Seiji Sugita and Peter H. Schultz, "Efficient Cyanide Formation Due to Impacts of Carbonaceous Bodies on a Planet with a Nitrogen-Rich Atmosphere," *Geophysical Research Letters* 36 (2009), doi:10.1029/2009GL040252. For the formation of organic compounds in comets and asteroids, see George D. Cody et al., "Establishing a Molecular Relationship Between Chondritic and Cometary Organic Solids," *PNAS* 108, no. 48 (November 29, 2011), www.pnas.org/cgi/doi/10.1073/pnas.1015913108 (accessed November 23, 2014).
20. Scott F. Gilbert, Jan Sapp, and Alfred I. Tauber, "A Symbiotic View of Life: We Have Never Been Individuals," *Quarterly Review of Biology* 87, no. 4 (December 2012): 325–41.
21. Friedrich Nietzsche, "On Truth and Lies in a Nonmoral Sense," in *The Nietzsche Reader*, ed. Keith Ansell Pearson and Duncan Large (Oxford: Blackwell, 2006), 114–23 (118).
22. Cary Wolfe, *What Is Posthumanism?* (Minneapolis: University of Minnesota Press, 2012).
23. *Oxford English Dictionary*, "gather," 4.a., b., c.; "glean," http://www.oed.com (accessed December 17, 2012): "1. To gather or pick up ears of corn which have been left by the reapers."
24. Luce Irigaray, *This Sex Which Is Not One*, trans. Catherine Porter and Carolyn Burke (Ithaca: Cornell University Press, 1985), 29.

25. Ibid., 26.
26. Plato, *Timaeus,* "Space" or sections 48e–53c. See Edward Casey, *The Fate of Place* (Berkeley: University of California Press, 1998).
27. Claude Lévi-Strauss, *Structural Anthropology* (New York: Basic Books, 1974), 4–5.
28. Alphonso Lingis, *The Imperative* (Bloomington: Indiana University Press, 1998), 68.
29. I am in accord with Heidegger here: *Being and Time*, 1.6.40.
30. Heidegger, *Being and Time*, 1.6 (180).
31. Jean-Luc Marion calls this a *saturated phenomenon*: *In Excess: Studies of Saturated Phenomena* (New York: Fordham University Press, 2002), 54–81.
32. Bill Brown, "Thing Theory," *Critical Inquiry* 28, no. 1 (Autumn 2001): 1–22.
33. Jacques Derrida, *Of Grammatology*, trans. Gayatri Spivak (Baltimore: Johns Hopkins University Press, 1998), 159.
34. Martin Heidegger, "Language," in *Poetry, Language, Thought*, trans. Albert Hofstadter (New York: Harper and Row, 1971), 187–210.
35. Derrida makes this clear in the introduction to *Of Grammatology*, 9.
36. Timothy Morton, *The Ecological Thought* (Cambridge: Harvard University Press, 2010), 8, 28–38.
37. Derrida, *Of Grammatology*, 70.
38. Ibid., 167.
39. Ibid., 92, 149, 222, 254, 287, 288, 299, 332, 348.
40. Ibid., 287.
41. Ibid., 299.
42. To this extent, Heidegger's meditation on technology (from which I borrow the idea of "worldview") is far too optimistically anachronistic. "The Question Concerning Technology," in Martin Heidegger, *Basic Writings: From "Being and Time" to "The Task of Thinking,"* ed. David Krell (New York: HarperCollins, 1993), 307–41.
43. Joseph Nesme, Sébastien Cécillon, Tom O. Delmont, Jean-Michel Monier, Timothy M. Vogel, and Pascal Simonet, "Large-Scale Metagenomic-Based Study of Antibiotic Resistance in the Environment," *Current Biology* 24, no. 10 (May 2014): 1096–100.
44. My thinking on *dialetheias* is deeply indebted to Graham Priest. See, for example, Graham Priest and Francesco Berto, "Dialetheism," in *The Stanford Encyclopedia of Philosophy* (Spring 2013 edition), ed. Edward N.

Zalta, http://plato.stanford.edu/archives/spr2013/entries/dialetheism/ (accessed May 14, 2015).

45. I am, of course, referring to Jamie Uys, dir., *The Gods Must Be Crazy* (Ster Kinekor and Twentieth Century Fox, 1980).
46. Michael Taussig, "Viscerality, Faith and Skepticism," in Birgit Meyer and Peter Pels, eds., *Magic and Modernity: Interfaces of Revelation and Concealment* (Stanford: Stanford University Press, 2003), 272–341 (273).
47. Mathias Guenther, *Tricksters and Trancers: Bushman Religion and Society* (Bloomington: Indiana University Press, 1999), 66.
48. Ibid., 226–47.
49. Robert Bellah, *Religion in Human Evolution: From the Paleolithic to the Axial Age* (Cambridge: Belknap, 2011), 150–53.
50. Jaynes, *The Origin of Consciousness*, 75–81, 84–94.
51. Ibid., 66, 84–85, 126–45.
52. Erwin Schrödinger, *What Is Life? The Physical Aspect of the Living Cell* in *What Is Life? With Mind and Matter and Autobiographical Sketches* (Cambridge: Cambridge University Press, 2012).
53. Aaron O'Connell, "Quantum Ground State and Single-Phonon Control of a Mechanical Resonator," *Nature* 464 (April 2010): 697–703.
54. Gilles Deleuze and Félix Guattari, *Anti-Oedipus: Capitalism and Schizophrenia*, trans. Robert Hurley, Mark Seem, and Helen R. Lane (Minneapolis: University of Minnesota Press, 1983), 122.
55. Immanuel Kant, *Critique of Pure Reason,* trans. Norman Kemp Smith (New York: St. Martin's, 1965), 84–85, A45–A47/B63–64.
56. Plato, *Phaedrus*, 55; 265e, http://classics.mit.edu/Plato/phaedrus.html (accessed April 30, 2015).
57. Heidegger, *Contributions to Philosophy (From Enowning),* trans. Parvis Emad and Kenneth Maly (Bloomington: Indiana University Press, 1999), 153.
58. Cary Wolfe, *Before the Law: Humans and Other Animals in a Biopolitical Frame* (Chicago: University of Chicago Press, 2013), 45–46. The bioart of Oron Catts (The Tissue Culture and Art Project, since 1996) has shown how artificial meat is synthesized.
59. Edmund Husserl, "Prolegomena to Pure Logic," in *Logical Investigations*, ed. Dermot Moran, trans. J. N. Findlay (London: Routledge, 2006), 1:1–161.
60. Henry Munn, "The Mushrooms of Language," in Michael J. Harner, ed., *Hallucinogens and Shamanism* (Oxford: Oxford University Press, 1973),

86–122. Terence McKenna, *Food of the Gods: The Search for the Original Tree of Knowledge: A Radical History of Plants, Drugs, and Human Evolution* (New York: Bantam, 1992). The scholarship of Dennis McKenna, brother of Terence, is an invaluable resource in this field.

61. Lin Zhang et al., "Exogenous Plant MIR168a Specifically Targets Mammalian LDLRAP1: Evidence of Cross-Kingdom Regulation by MicroRNA," *Cell Research* (2012): 107–26.
62. Jeffrey Kripal, *Authors of the Impossible: The Paranormal and the Sacred* (Chicago: University of Chicago Press, 2010), 162, 188.
63. See, for instance, Nicholas Royle's magnificent *Telepathy and Literature: Essays on the Reading Mind* (Oxford: Blackwell, 1991).
64. Bron Taylor, *Dark Green Religion: Nature Spirituality and the Planetary Future* (Berkeley: University of California Press, 2010).
65. I have been deeply inspired throughout this project by Nicholas Royle, *Veering: A Theory of Literature* (Edinburgh: Edinburgh University Press, 2011); the citation is from 2–5.
66. Alexei Sharov and Richard Gordon, "Life Before Earth" (2013), arXiv:1304.3381.
67. Jeremy L. England, "Statistical Physics of Self-Replication," *Journal of Chemical Physics* 139, 121923 (2013), doi: 10.1063/1.4818538.
68. Stanley M. Awramik and Kathleen Grey, "Stromatolites: Biogenicity, Biosignatures, and Bioconfusion," in Richard B. Hoover, Gilbert V. Levin, Alexei Y. Rozanov, and G. Randall Gladstone, eds., *Astrobiology and Planetary Missions,* proceedings of the SPIE 5906 (2005), doi: 10.1117/12.625556.
69. George Johnson, "A Tumor, the Embryo's Evil Twin," *New York Times,* March 17, 2014, http://www.nytimes.com/2014/03/18/science/a-tumor-the-embryos-evil-twin.html?_r=0 (accessed March 21, 2014).
70. David Chalmers, *The Conscious Mind: In Search of a Fundamental Theory* (Oxford: Oxford University Press, 1996); Galen Strawson, *Mental Reality* (Cambridge: MIT Press, 2010).
71. The Marx Brothers (Leo McCarey, dir.), *Duck Soup* (Paramount Pictures, 1933).
72. Elizabeth Lunbeck, *The Americanization of Naricissism* (Cambridge: Harvard University Press, 2014).
73. Jacques Derrida, *Voice and Phenomenon: Introduction to the Problem of the Sign in Husserl's Phenomenology,* trans. Edward Lawlor (Evanston: Northwestern University Press, 2011), 56, 67; *On Touching—Jean-Luc*

Nancy, trans. Christine Irizarry (Stanford: Stanford University Press 2005), 246–47.

74. Georg Wilhelm Friedrich Hegel, *The Science of Logic,* trans. George di Giovanni (New York: Cambridge University Press, 2010), 337–505.
75. Georg Wilhelm Friedrich Hegel, *Hegel's Phenomenology of Spirit,* trans. A. V. Miller, analysis and foreword by J. N. Findlay (Oxford: Oxford University Press, 1977), 9.
76. Lunbeck, *The Americanization of Narcissism,* 271.
77. Schopenhauer, *The World as Will and Representation,* 1:137n13, §26.
78. Stephen Messenger, "Extinct Tree Grows from Ancient Jar of Seeds Unearthed by Archaeologists," *Treehugger,* October 5, 2013, http://www.treehugger.com/natural-sciences/extinct-tree-grows-anew-after-archaeologists-dig-ancient-seed-stockpile.html (accessed October 7, 2013).
79. John Milton, *Paradise Lost,* ed. Alastair Fowler (London: Longman, 1971).
80. Jacques Lacan, *Le séminaire, Livre III: Les psychoses* (Paris: Seuil, 1981), 48.
81. Friedrich Nietzsche, *Thus Spoke Zarathustra* (Cambridge: Cambridge University Press, 2006), 6.
82. René Descartes, *Meditations on First Philosophy* (Indianapolis: Hackett, 1993), 16–17.
83. See Daniel Chamovitz, *What a Plant Knows: A Field Guide to the Senses* (New York: Farrar, Straus and Giroux, 2013).
84. T. S. Eliot, *Burnt Norton, Four Quartets* (Orlando: Mariner, 1968), ll. 28–29.
85. Schopenhauer, *The World as Will and Representation,* 1.3.201.
86. Immanuel Kant, *Critique of Judgment,* trans. Werner Pluhar (Indianapolis: Hackett, 1987), §58, 221–22.
87. Schopenhauer, *The World as Will and Representation,* 1.3.222.
88. Ibid., 1.3.182.
89. Charles Darwin, *The Descent of Man* (London: Penguin, 2004), 114–16.
90. Ibid., 241–49.
91. Joan Roughgarden, *Evolution's Rainbow: Diversity, Gender, and Sexuality in Nature and People* (Berkeley: University of California Press, 2009), 30.
92. I use Judith Roof's invaluable concept of DNA as cipher rather than as code. Judith Roof, *The Poetics of DNA* (Minneapolis: University of Minnesota Press, 2007), 78, 81–82.
93. Kant, *Critique of Judgment,* 49.
94. Jacques Derrida, "There Is No One Narcissism: Autobiophotographies," in *Points: Interviews, 1974–1994,* ed. Elisabeth Weber, trans. Peggy Kamuf et al. (Stanford: Stanford University Press, 1995), 196–215 (199).

95. The Pascal quotation forms one of the epigraphs to Emmanuel Levinas, *Otherwise Than Being: Or Beyond Essence,* trans. Alphonso Lingis (Pittsburgh: Duquesne University Press, 1998), vii. The most profound discussion of this is found in Emmanuel Levinas, *Totality and Infinity: An Essay on Exteriority,* trans. Alphonso Lingis (Pittsburgh: Duquesne University Press, 1969), 37–38. See also Emmanuel Levinas, "Interview with François Piorié," in *Is It Righteous to Be? Interviews with Emmanuel Levinas,* ed. Jill Robbins (Stanford: Stanford University Press, 2001), 23–83 (53).
96. Peter Atterton, "Do I Have the Right to Be?" *New York Times,* July 5, 2014, http://opinionator.blogs.nytimes.com/2014/07/05/do-i-have-the-right-to-be/ (accessed July 7, 2014).
97. Hegel, *Hegel's Phenomenology of Spirit,* 9.
98. Georg Wilhelm Friedrich Hegel, *The Philosophy of History,* trans. J. Sibree (Mineola: Dover, 2004), 8–79.
99. George Dunning, dir., *Yellow Submarine* (Apple and United, 1968).
100. Douglas Hofstadter, *Gödel, Escher, Bach: An Eternal Golden Braid* (New York: Basic Books, 1999), 76.
101. Paul Tillich, *Systematic Theology 1* (Chicago: University of Chicago Press, 1951), 188.
102. For much further detail on this topic, see David Macauley, *Elemental Philosophy: Earth, Air, Fire, and Water as Environmental Ideas* (Albany: SUNY Press, 2010).
103. Tillich, *Systematic Theology 1,* 188.
104. Jane Taylor, "The Star," in *Rhymes for the Nursery* (London, 1806), 10.
105. L. Pustil'nik and G. Yom Din, "On Possible Influence of Space Weather on Agricultural Markets: Necessary Conditions and Probable Scenarios," *Astrophysical Bulletin* 68, no. 1 (2013): 1–18.
106. Avicenna, *Metaphysics,* I.8, 53.13–15. I quote the commonly cited version, which appears to be a translation of *La Métaphysique du Shifā Livres I à V,* ed. Georges Anawati (Paris: Vrin, 1978), cited for instance in Laurence R. Horn, "Contradiction," in *The Stanford Encycopedia of Philosophy* (spring 2014 edition), ed. Edward N. Zalta, http://plato.stanford.edu/archives/spr2014/entries/contradiction/ (accessed July 15, 2015). The more readily available and recent English translation is by Michael E. Marmura (Provo: Brigham Young University Press, 2005), 43.
107. The Doors, "The Celebration of the Lizard," in *Absolutely Live* (Elektra, 1970). The Beatles, "I Am the Walrus," in *Magical Mystery Tour* (EMI, 1967). George W. Bush, press conference, April 18, 2006, https://youtu.be

/irMeHmlxE9s (accessed July 23, 2015). The parodic mash-up of Bush and the Beatles is available at https://youtu.be/1ks3ljVr9ZA (accessed July 23, 2015).

108. Henri Bergson, "Laughter," in Wylie Sypher, ed. and intro., *Comedy: "An Essay on Comedy" by George Meredith and "Laughter" by Henri Bergson* (Baltimore: Johns Hopkins University Press, 1956), 59–190.

THE THIRD THREAD

1. Charles Baudelaire, "Spleen," in *Les Fleurs du Mal*, trans. Richard Howard (Brighton: Harvester, 1982), 76.
2. Graham Priest, "The Logic of Buddhist Philosophy," *Aeon* 2014, http://aeon.co/magazine/world-views/logic-of-buddhist-philosophy/ (accessed April 30, 2015).
3. "This Cold Snap Is Making It Colder than the Surface of Mars," Smithsonian Smart News, January 2, 2014, http://blogs.smithsonianmag.com/smartnews/2014/01/this-cold-snap-is-making-it-colder-than-the-surface-of-mars/ (accessed January 8, 2014).
4. Lisa A. Schulte Moore, "Prairie Strips: Bringing Biodiversity, Improved Soil Quality, and Soil Protection to Agriculture," *Missouri Prairie Journal* 35 (2014): 12–15.
5. Takao Furuno, "The Power of Duck: Integrated Rice and Duck Farming," np. Shaikh Tanveer Hossain et al., "Effect of Integrated Rice-Duck Farming on Rice Yield, Farm Productivity, and Rice-Provisioning Ability of Farmers," *Asian Journal of Agriculture and Development* 2, no. 1 (1992): 79–86.
6. S. Jose, A. R. Gillespie, and S. G. Pallardy, "Interspecific Interactions in Temperate Agroforestry," *Agroforestry Systems* 61 (2004): 237–55.
7. H. F. van Emden, "Conservation Biological Control: From Theory to Practice," in Pedro Barbosa, ed., *Conservation Biological Control* (San Diego: Harcourt Brace, 1998), 199–208.
8. A. McLeod et al., " 'Beetle Banks' as Refuges for Beneficial Arthropods in Farmland: Long-Term Changes in Predator Communities and Habitat," *Agricultural and Forest Entomology* 6 (2004): 147–54.
9. "Alberta Artist Copyrights Land as Artwork to Keep Oil Companies at Bay," *Cantech Letter*, May 27, 2014, http://www.cantechletter.com/2014/05/alberta-artist-copyrights-land-artwork-keep-oil-companies-bay/ (accessed June 11, 2014).

10. Jón Gnarr, *Gnarr! How I Became the Mayor of a Large City in Iceland and Changed the World*, trans. Andrew Brown (New York: Melville House, 2014).
11. Mathias Guenther, *Tricksters and Trancers: Bushman Religion and Society* (Indianapolis: Indiana University Press, 1999).
12. David Graeber, "What's the Point if We Can't Have Fun?" *Baffler* 24 (2014), http://thebaffler.com/past/whats_the_point_if_we_cant_have_fun (accessed February 15, 2014).
13. Fredy Perlman, "Against His-Story, Against Civilization!" in John Zerzan, ed., *Against Civilization: Readings and Reflections* (Port Townsend, WA: Feral House, 2005), 27–30 (28).
14. Gregory Bateson, "A Theory of Play and Fantasy," in *Steps to an Ecology of Mind*, foreword Mary Catherine Bateson (Chicago: University of Chicago Press, 2000), 177–93.
15. Friedrich Schiller, *On the Aesthetic Education of Man In a Series of Letters, English and German Facing*, ed. and trans. with intro., commentary, and glossary by Elizabeth M. Wilkinson and L. A. Willoughby (Oxford: Clarendon, 1983), 107.
16. I am grateful to Federico Campagna for this asymmetrical chiasmus.
17. Situationist slogan.
18. Roy Scranton, "Learning How to Die in the Anthropocene," *New York Times*, November 10, 2013, http://opinionator.blogs.nytimes.com/2013/11/10/learning-how-to-die-in-the-anthropocene/?_r=1&&pagewanted=print (accessed November 11, 2013).
19. See page 36.
20. Sigmund Freud, *The Uncanny*, trans. David McClintock and Hugh Haughton (London: Penguin, 2003).
21. René Descartes, *Meditations and Other Metaphysical Writings*, trans. and intro. Desmond M. Clarke (London: Penguin, 2000).
22. Freud, *The Uncanny*. The Cure, "A Forest," in *Seventeen Seconds* (Elektra/Asylum, 1980).
23. Jonathan Rottenberg, "The Depression Epidemic Will Not Be Televised," *Huffington Post*, May 22, 2014, http://www.huffingtonpost.com/jonathan-rottenberg/the-depression-epidemic-will-not-be-televised_b_5367479.html (accessed November 10, 2014).
24. Colin Campbell, *The Romantic Ethic and the Spirit of Modern Consumerism* (New York: Basil Blackwell, 1987); "Understanding Traditional and Modern Patterns of Consumption in Eighteenth-Century England: A Character-

Action Approach," in John Brewer and Roy Porter, eds., *Consumption and the World of Goods* (New York: Routledge, 1993), 40–57.

25. Cary Wolfe, *Animal Rites: American Culture, the Discourse of Species, and Posthumanist Theory* (Chicago: University of Chicago Press, 2003), 6.
26. Julian Jaynes, *The Origin of Consciousness in the Breakdown of the Bicameral Mind* (New York: Mariner, 2000).
27. Jeffrey Kripal, *The Serpent's Gift: Gnostic Reflections on the Study of Religion* (Chicago: University of Chicago Press, 2006), 6–7.
28. Alphonso Lingis, *The Imperative* (Bloomington: Indiana University Press, 1998), 2.
29. Kate Soper, "Alternative Hedonism, Cultural Theory and the Role of Aesthetic Revisioning," *Cultural Studies* 22, no. 5 (September 2008): 567–87.
30. Jean Antheleme Brillat-Savarin, *The Physiology of Taste*, trans. Anne Drayton (Harmondsworth: Penguin, 1970), 13. Ludwig Feuerbach, *Gesammelte Werke II, Kleinere Schriften* ed. Werner Schuffenhauer (Berlin: Akadamie, 1972), 4.27.
31. Luce Irigaray, *This Sex Which Is Not One,* trans. Catherine Porter and Carolyn Burke (Ithaca: Cornell University Press, 1985), 88–96.
32. Lingis, *The Imperative,* 38.
33. A fact related intimately to the history of definitions of homosexuality. If desiring a thing is inherently perverse because (1) there's no accounting for taste and (2) the thing seduces you, there is something "wrong" with consumerism as such. This correlates closely with the history of addiction speech. See Eve Sedgwick, "Epidemics of the Will," in *Tendencies* (Durham: Duke University Press, 1993), 130–42.
34. Sigmund Freud, "The Ego and the Id," in *The Standard Edition of the Complete Psychological Works of Sigmund Freud* (New York: Norton, 1990), 19:24.
35. See Annie Sprinkle and Elizabeth Stephens, "Sexecology: Where Art Meets Practice Meets Activism," http://sexecology.org (accessed May 16, 2015).
36. Donna Haraway, "The Promises of Monsters: A Regenerative Politics for Inappropriate/d Others, in Lawrence Grossberg, Cary Nelson, Paula A. Treichler, eds., *Cultural Studies* (New York; Routledge, 1992), 295–337.
37. Georg Wilhelm Friedrich Hegel, *Hegel's Phenomenology of Spirit,* trans. A. V. Miller (Oxford: Oxford University Press, 1977), 383–409.
38. Donald Braman, Dan M. Kahan, Maggie Wittlin, and Paul Slovic, "The Polarizing Impact of Science Literacy and Numeracy on Perceived Climate Change Risks," *Nature Climate Change* 732 (2012).

39. Simon Keller, "Empathizing with Skepticism about Climate Change," paper presented to the Cultures of Energy Seminar, Rice University, December 5, 2012.
40. See, for example, William Jordan, *The Sunflower Forest: Ecological Community and the New Communion with Nature* (Berkeley: University of California Press, 2003).
41. For an extensive exploration, see Ruth Leys, *From Guilt to Shame: Auschwitz and After* (Princeton: Princeton University Press, 2009).
42. Julia Kristeva, *Powers of Horror: An Essay on Abjection,* trans. Leon S. Roudiez (New York: Columbia University Press, 1982), 5–6.
43. See page 78.
44. Like Irigaray's air. See Luce Irigaray, *The Forgetting of Air in Martin Heidegger,* trans. Mary Beth Mader (Austin: University of Texas Press, 1999).
45. Kristeva, *Powers of Horror.*
46. See Jean-Luc Marion, *In Excess: Studies of Saturated Phenomena,* trans. Robyn Horner and Vincent Berraud (New York: Fordham University Press, 2010).
47. Pink Floyd, "Welcome to the Machine," in *Wish You Were Here* (EMI, 1975).
48. According to some interpretations of patches of energy in the Cosmic Microwave Background: Stephen M. Feeney et al., "First Observational Tests of Eternal Inflation: Analysis Methods and WMAP 7-Year Results," *Physical Review* D 84, no. 4 (2011), DOI: 10.1103/PhysRevD.84.043507. Sigmund Freud, "Abandoned Object Cathexes," in *The Ego and the Id,* trans. Joan Riviere, revised and ed. James Strachey, intro. Peter Gay (New York: Norton 1989), 24.
49. One of the best meditations on this is Dylan Trigg, *The Thing: A Phenomenology of Horror* (Alresford: Zero, 2014).
50. Ray Brassier, *Nihil Unbound: Enlightenment and Extinction* (New York: Palgrave Macmillan, 2007), 48.
51. Masahiro Mori, "The Uncanny Valley" (*Bukimi no tani*), trans. K. F. MacDorman and T. Minato, *Energy* 7, no. 4 (1970): 33–35.
52. Giorgio Agamben, *The Open: Man and Animal,* trans. Kevin Attell (Stanford: Stanford University Press, 2004), 33–38.
53. Michel Foucault, *"Society Must Be Defended": Lectures at the Collège de France, 1975–1976,* trans. David Macey (New York: Picador, 2003), 243–247.
54. See, for instance, Luc Ferry, *The New Ecological Order,* trans. Carol Volk (Chicago: University of Chicago Press, 1995).

55. Eugene Thacker, *In the Dust of This Planet: Horror of Philosophy 1* (Alresford: Zero, 2011), 104.
56. It is not a coincidence, argues Graham Harman, that the first four philosophers of speculative realism are fans of Lovecraft. Exemplary studies from this lineage include Graham Harman's *Weird Realism: Lovecraft and Philosophy* (Alresford: Zero, 2012) and Thacker, *In the Dust of This Planet* (esp. pp. 9, 19, 74–84, and 104).
57. Jean-Pierre Jeunet, dir., *Alien Resurrection* (Twentieth Century Fox, 1997).
58. Brassier, *Nihil Unbound,* xi.
59. Ibid., 239.
60. Lars von Trier, dir., *Melancholia* (Dallas: Magnolia Pictures, 2011).
61. This is the view of Quentin Meillassoux; see Rick Dolphijn and Iris van der Tuin, *New Materialisms: Interviews and Cartographies* (Ann Arbor: Open Humanities, 2012), 71.
62. Henri Bergson, *Laughter: An Essay on the Meaning of the Comic,* trans. Cloudesley Brereton (New York: Macmillan, 1911), 10.
63. See, for example, Frances Moore Lappé, *Diet for a Small Planet* (New York: Ballantine, 1971).
64. David Holmgren, "Crash on Demand: Welcome to the Brown Tech Future," http://holmgren.com.au/wp-content/uploads/2014/01/Crash-on-demand.pdf (accessed November 26, 2014). "Crash on Demand: A Concise Version," http://holmgren.com.au/wp-content/uploads/2014/02/cons-COD.pdf (accessed November 26, 2014).
65. Hermann Scheer, *The Solar Economy: Renewable Energy and a Sustainable Future* (New York: Routledge, 2004).
66. United Nations Conference on Trade and Development, "Wake Up Before It Is Too Late: Make Agriculture Truly Sustainable Now for Food Security in a Changing Climate," *Trade and Environment Review 2013* (New York: United Nations, 2013).
67. Theodor W. Adorno, *Aesthetic Theory,* ed. and trans. Robert Hullot-Kentor (Minneapolis: University of Minnesota Press, 1997), 331.
68. As many artists have noted. See, for instance, http://collectingseminar.wordpress.com/2008/11/02/claes-oldenburgs-the-ray-gun-wing/ (accessed July 17, 2015).
69. Bateson, "A Theory of Play and Fantasy," 182.
70. Shelley, "A Defence of Poetry," in *Shelley's Poetry and Prose* (New York: Norton, 2002), 530.
71. John Carpenter, dir., *The Thing* (Universal Studios, 1982).

72. Ridley Scott, dir., *Blade Runner* (Warner Bros., 1982).
73. A substantial path has been mapped out in Daniel Tiffany, *My Silver Planet: A Secret History of Poetry and Kitsch* (Baltimore: Johns Hopkins University Press, 2014). I am thinking of the traditional critiques of kitsch, above all by Clement Greenberg and Adorno.
74. Andrew Stanton, dir., *Wall•E* (Pixar Animation Studios, 2008).
75. Jane Bennett, "Vitality and Self-Interest," in *Vibrant Matter: A Political Ecology of Things* (Durham: Duke University Press, 2010), 110–22.
76. Lewis Carroll, *Alice Through the Looking Glass* in *The Annotated Alice: The Definitive Edition,* ed. Martin Gardner (New York: Norton, 2000), 157.
77. Syd Barrett, "Jugband Blues," in Pink Floyd, *A Saucerful of Secrets* (EMI, 1968).
78. William Shakespeare, *Twelfth Night,* ed. Elizabeth Story Donno (Cambridge: Cambridge University Press, 1994).
79. Nicholas Royle, *Veering: A Theory of Literature* (Edinburgh: Edinburgh University Press, 2011), 5.
80. Judea Pearl, *Causality: Models, Reasoning, and Inference* (Cambridge: Cambridge University Press, 2010), 78–85.
81. Aaron O'Connell et al., "Quantum Ground State and Single-Phonon Control of a Mechanical Resonator," *Nature* 464 (April 2010): 697–703.
82. Again, see Marion, *In Excess.*
83. Lingis, *The Imperative*, 159–63.
84. William Blake, "The Human Abstract," in *The Complete Poetry and Prose of William Blake,* ed. David V. Erdman (New York: Doubleday, 1988), ll. 1–2. For idiot compassion, see Chögyam Trungpa, *Training the Mind and Cultivating Loving-Kindness,* ed. Judith Lief (Boston: Shambhala, 1993), 18–19.
85. See Trebbe Johnson's Radical Joy for Hard Times, http://www.radical joyforhardtimes.org (accessed May 16, 2015).
86. Cited in Lindsay Holmes, "Six Things Not to Say to Someone with Depression," *Huffington Post,* January 29, 2014, http://www.huffingtonpost .com/2014/01/29/what-not-to-say-to-someon_n_4675854.html (accessed April 30, 2015).
87. Meredith Melnick, "Robin Williams's Death Reveals How Hard It Can Be to Climb Out of Depression," *Huffington Post,* August 11, 2014, http://www .huffingtonpost.com/2014/08/11/robin-williams-depression_n_5670256 .html (accessed November 10, 2014).

88. Gregory Bateson, "The Cybernetics of 'Self': A Theory of Alcoholism," in *Steps to an Ecology of Mind*, 309–37 (320–22).
89. Kripal, *The Serpent's Gift*, 125.
90. Irigaray, *This Sex Which Is Not One,* 101. The feminist adaptation of Plato's chora is performed by Julia Kristeva, *Revolution in Poetic Language* (New York: Columbia University Press, 1984), 25–42.
91. James Cameron, dir. *Avatar* (Twentieth Century Fox, 2009).
92. Theodor Adorno, *Prisms* (Cambridge: MIT Press, 1983), 72.
93. Robert Bellah, *Religion in Evolution: From the Paleolithic to the Axial Age* (Cambridge: Belknap, 2011), ix–xiv.
94. See Tim Ingold, "Footprints Through the Weather-World: Walking, Breathing, Knowing," *Journal of the Royal Anthropological Institute* 16 (May 2010): 121–39.
95. See, for example, Glen Mazis, *Humans, Animals, Machines: Blurring Boundaries* (Albany: SUNY Press, 2008).
96. I prefer this to the image of the flayed man in Michel Serres, *The Five Senses: A Philosophy of Mingled Bodies,* trans. Margaret Sankey and Peter Crowley (New York: Continuum, 2009), viii.
97. Irigaray, *This Sex Which Is Not One,* 31.
98. Deleuze's concept of smoothness could come into play here: something so granular that we lose track of detail. See, for example, *The Fold: Leibniz and the Baroque,* trans. Tom Conley (Minneapolis: University of Minnesota Press, 1993). I have also benefited from Kali Rubaii, "In the Path of the Witness-Perpetrator: Concrete and Chemicals in Anbar, Iraq," paper given at Ethics, Agency and Aesthetics in the Anthropocene: A Symposium, UC Berkeley, April 17, 2015.
99. William Blake, "Auguries of Innocence," in *The Complete Poetry and Prose of William Blake,* ll. 111–12.

ENDING BEFORE THE BEGINNING

1. http://infinityburialproject.com (accessed May 12, 2015).

INDEX

PREVIOUSLY PUBLISHED WELLEK LIBRARY LECTURES

The Breaking of the Vessels (1983)
HAROLD BLOOM

In the Tracks of Historical Materialism (1984)
PERRY ANDERSON

Forms of Attention (1985)
FRANK KERMODE

Memoires for Paul de Man (1986)
JACQUES DERRIDA

The Ethics of Reading (1987)
J. HILLIS MILLER

Peregrinations: Law, Form, Event (1988)
JEAN-FRANÇOIS LYOTARD

A Reopening of Closure: Organicism Against Itself (1989)
MURRAY KRIEGER

Musical Elaborations (1991)
EDWARD W. SAID

Three Steps on the Ladder of Writing (1993)
HÉLÈNE CIXOUS

The Seeds of Time (1994)
FREDRIC JAMESON

Refiguring Life: Metaphors of Twentieth-Century Biology (1995)
EVELYN FOX KELLER

The Fateful Question of Culture (1997)
GEOFFREY HARTMAN

The Range of Interpretation (2000)
WOLFGANG ISER

History's Disquiet: Modernity, Cultural Practice, and the Question of Everyday Life (2000)
HARRY HAROOTUNIAN

Antigone's Claim: Kinship Between Life and Death (2000)
JUDITH BUTLER

The Vital Illusion (2000)
JEAN BAUDRILLARD

Death of a Discipline (2003)
GAYATRI CHAKRAVORTY SPIVAK

Postcolonial Melancholia (2005)
PAUL GILROY

On Suicide Bombing (2007)
TALAL ASAD

Cosmopolitanism and the Geographies of Freedom (2009)
DAVID HARVEY

Hiroshima After Iraq: Three Studies in Art and War (2009)
ROSALYN DEUTSCHE

Globalectics: Theory and the Politics of Knowing (2012)
NGUGI WA THIONG'O

Violence and Civility: On the Limits of Political Philosophy (2015)
ÉTIENNE BALIBAR